ENVIRONMENTAL SCIENCE, ENGINEERING AND TECHNOLOGY

WOODLANDS

STRUCTURE, SPECIES DIVERSITY AND SUSTAINABLE MANAGEMENT

Environmental Science, Engineering and Technology

Additional books in this series can be found on Nova's website under the Series tab.

Additional e-books in this series can be found on Nova's website under the e-book tab.

ENVIRONMENTAL SCIENCE, ENGINEERING AND TECHNOLOGY

WOODLANDS

STRUCTURE, SPECIES DIVERSITY AND SUSTAINABLE MANAGEMENT

LAZARO MANZANARES
EDITOR

nova publishers
New York

For permission to use material from this book please contact us:
Telephone 631-231-7269; Fax 631-231-8175
Web Site: http://www.novapublishers.com

Additional color graphics may be available in the e-book version of this book.

Library of Congress Cataloging-in-Publication Data

ISBN: 978-1-62618-177-9

Published by Nova Science Publishers, Inc. † New York

CONTENTS

PREFACE

Forest and woodland ecosystems in the world serve important ecological functions and also contribute to the economic, aesthetic and spiritual health of humans. In this book, the authors discuss the structure, species diversity and sustainable management of woodlands across the globe. Topics include changes in composition and threats to the sustainable management of woodlands in Portugal; the role and diversity of soil fauna in woodlands; climate change and forest fires in the Ave Region of Northwest Portugal; drivers of deforestation and the potential for carbon trading in the Miombo woodlands of Zambia; soil microbial diversity in the dry woodlands of Central-Western Argentina; and developing silvicultural systems based on disturbance-recovery knowledge of the Southern African Miombo woodlands.

Chapter 1 – Since the beginning of the last century, important changes have occurred in the woodlands of mainland Portugal, which have almost tripled in size, occupying around 40% of the territory according the most recent statistics. Such changes are mainly supported by an increase in the area occupied by fast-growing species (e.g. *Pinus pinaster*), some of which are exotic, such as *Eucalyptus globulus*. These changes are of special significance north of the River Tejo, where the native forests are severely reduced to small highly disturbed patches. In the southern part of mainland Portugal, the changes are less important, as the native forests still maintain an important presence in the landscape, despite deep structural changes. This work describes the diversity of the native forests and their potential distribution. It also discusses the different threats posed to sustainable management, such as forest fires, epidemic diseases and invasion by alien species.

Chapter 2 – The increasing anthropic pressure on the environment is leading, in most parts of the world, to a rapid change in land use and an intensification of agricultural activities, in many cases at the expense of forest ecosystems. These processes often result in soil degradation and a consequent loss of soil quality and soil biodiversity. Soil is one of the most important reservoirs of biodiversity. Within it is found the habitat of the roots of plants, microorganisms such as viruses, bacteria, fungi and algae and numerous invertebrates and vertebrates. It reflects ecosystem metabolism since all the bio-geo-chemical processes of the different ecosystem components are combined within it. Organic matter degradation and nutrient cycling in the belowground ecosystems are regulated by very complex biotic and abiotic interactions, where soil microflora and soil fauna directly and indirectly affect soil functioning. Soil fauna play an essential role in several soil ecosystem functions; furthermore, it is often used to provide soil quality indicators. In woodland ecosystems, plant composition, bearing the characteristics of the litter and the rhizosphere, affects both edaphic community structure and the dynamics of decomposition and recycling. On the other hand, edaphic community structure affects the decomposition of organic matter and the formation of the humus profile. In fact, it has been shown that soil invertebrates improve the soil fertility and productivity of forest ecosystems.

Chapter 3 – There is a wide array of studies and evidence that climate is changing and these changes will manifest themselves very differently in various areas of the planet. The weather conditions that occur in Portugal, especially during the summer, are favorable to the occurrence of forest fires. However, the ignition and spread of a fire depends on the interaction of several factors besides the weather, namely the presence of fuel and rugged terrain. It is expected that fire regimes immediately respond to climate change, and may even outweigh the direct effects of global warming in the patterns of species distribution and productivity. The project ADAPTACLIMA - Adaptation to the effects from climate change (INTERREG SUDOE) was based on the preparation of a series of studies on forecasting and analyzing the vulnerabilities and potentialities in Southwest European territories, with the aim of creating a collaborative network of stable institutions. The purpose was to permit both the transmission of knowledge and exchange of experiences among members of the partnership, as well as mutual learning and co-generation of new knowledge. The fundamental task of the network was the preparation of a Plan for Adaptation to Climate Change in Space SUDOE to be implemented in the participating areas. Among the challenges facing SUDOE to mitigate climate change, two of them represent the strategic objectives of the project ADAPTACLIMA: on the one hand, to alert the populations of the SUDOE space to the real consequences of climate change; and, secondly, to promote and develop measures which help adapt society to future scenarios derived from these changes. In the northwest of Portugal and in particular in the Averegion, one of the main impacts expected from climate change is an increase in number and size of fires and their recurrence. In terms of plant species, those better adapted to fire will dominate, generating monospecific formations or small variations, wherein all the plant species are of the same age. Climate change may cause a substantial increase in the risk of fire. In addition, the period of fire occurrence will extend throughout the year, implying a larger fire-fighting organizational structure, which will maintain higher levels of alert for longer periods each year.

Chapter 4 – Africa's Miombo woodlands are vast, of high conservation value and essential for human well-being. This forest type plays a pivotal role in livelihoods of the majority of resource-poor rural communities in southern Africa. Majority of these communities have limited access to sources of employment and domestic income because they have little cash or other resources. It is, however, unfortunate that livelihood factors such as agricultural production and use of fire have constantly altered miombo woodlands. These livelihood factors have all profoundly contributed to degradation and in some cases depletion of the forest resources in southern Africa. Many studies have demonstrated that deforestation in miombo woodlands is one of the major contributors to global climate change. However, the question of what exactly drives woodlands deforestation at the regional scale remains largely unanswered. Over the last three decades, significant amount of evidence has coalesced around the general assumption that factors that drive tropical deforestation interact in a complex way and fluctuate significantly at global, regional and country levels. Not surprising all UNFCCC negotiations and key debates have tended to revolve around forests. This is mainly because of the realization that the objectives of mitigating climate change may not be reached without extensive action on deforestation and forest degradation. Therefore, a review of drivers of deforestation in miombo woodlands will help in deepening the authors' understanding of what drives miombo woodlands deforestation and what the potential for future carbon trade is like.

Chapter 5 – The dry woodlands cover a great area of central-western of Argentina, characterized by an E-W precipitation gradient ranging from 500 to 100 mm. Accordingly, two arid eco-regions with different dry woodland type are defined: Arid Chaco and Monte. Considerable researches have been done on biodiversity comparison between both eco-regions focusing on animal and vegetation communities, whereas the microbial community diversity has been neglected. However, microorganisms play a key role in the organic matter decomposition and nutrient availability, so that the microbial study increases current understanding of the ecosystem functionality. The authors analized bacteria and actinomycetes diversity of soil and litter in two dry woodlands of central-western Argentina and its response to ecological conditions, including seasonal climatic changes. The study sites were Chancaní Reserve in Arid Chaco woodland (31°22'S and 65°27'W) and Ñacuñan Reserve in Monte woodland (34º03'S and 67º58'W). At each site, three composite samples of soil (0-20 cm deep) and surface litter (0.16 m^2) were collected in wet and dry season. Genetic diversity by polymerase chain reaction-denaturing gradient gel electrophoresis (PCR-DGGE) of bacteria and actinomycete communities were performed and interpreted in terms of ribotype richness and frequency and similarity index. The DGGE profile analysis reveals 39 bacteria ribotypes and 31 actinomyceteribotypes, among them 20.5% and 29% respectively, were observed only once. Two actinomyceteribotypes showed 100% of frequency (present in all samples). The ribotype richness of bacteria and actinomycetes were similar in all samples, but actinomycete richness was different between soil and litter (16.75 vs. 9.75, respectively). The low similarity index (<0.80) in bacteria and actinomycete communities indicate a marked microbial diversity in dry woodlands of central-western Argentina. The authors concluded that a) there are great bacteria and actinomycete diversity in dry woodlands of central-western Argentina, and b) the structure of microbial community responds to ecological conditions. To the authors' knowledge, this is the first study that analizes bacteria and actinomycte diversity in a comparative way between dry woodland from realistic ecological conditions.

Chapter 6 – The current negative perception that various land use systems within the Southern African Miombo woodlands cause the loss of Miombo woodland and degradation of the environment influenced the development of forest management programmes in the Miombo ecoregion. Such programmes exclude many forms of woodland utilization because the available silvicultural systems are not aligned with the ecological characteristics of Miombo woodland ecosystems or their targeted species. This chapter explores the concept of disturbance-recovery processes, with recovery development as the basis for developing silvicultural systems suitable for managing Miombo woodland complexity. The size class profiles derived from various data sources are used to determine specific shade-or fire-tolerance characteristics of key economic and ecological tree species of Miombo woodlands. Additionally, the stand level grain analysis gives an indication of the scale of disturbances that drives Miombo woodland characteristics. Such information underlines the development and implementation of appropriate silvicultural management systems to ensure sustainable resource use from Miombo woodlands.

In: Woodlands
Editor: Lazaro Manzanares
ISBN: 978-1-62618-177-9

Chapter 1

CHANGES IN COMPOSITION AND THREATS TO THE SUSTAINABLE MANAGEMENT OF WOODLANDS IN PORTUGAL

Adélia Nunes*[*]*, Albano Figueiredo and António Campar de Almeida
Centro de Estudos em Geografia e Ordenamento do Território (CEGOT),
Departamento de Geografia da Faculdade de Letras,
Universidade de Coimbra, Coimbra, Portugal

ABSTRACT

Since the beginning of the last century, important changes have occurred in the woodlands of mainland Portugal, which have almost tripled in size, occupying around 40% of the territory according the most recent statistics. Such changes are mainly supported by an increase in the area occupied by fast-growing species (e.g. *Pinus pinaster*), some of which are exotic, such as *Eucalyptus globulus*. These changes are of special significance north of the River Tejo, where the native forests are severely reduced to small highly disturbed patches. In the southern part of mainland Portugal, the changes are less important, as the native forests still maintain an important presence in the landscape, despite deep structural changes.

[*] Tel.: +351 239 859900; Fax: +351 239 836733; *Corresponding author: adelia.nunes@ ci.uc.pt.

This work describes the diversity of the native forests and their potential distribution. It also discusses the different threats posed to sustainable management, such as forest fires, epidemic diseases and invasion by alien species.

INTRODUCTION

Forest and woodland ecosystems serve important ecological functions and also contribute to human economic, aesthetic and spiritual health. Mediterranean forests provide a wide array of environmental services and products. Watershed protection, landscape quality, soil conservation, carbon sequestration and recreation resources are among some of the public goods and externalities provided by Mediterranean forests that are seldom recognised and are difficult to evaluate and price (Merlo and Rojas Briales 2000, Croitoru 2007). Their conservation and appropriate management is crucial for the sustainability of the region's most strategic resources: soil and water. Moreover, Mediterranean forests provide wood products for timber and energy as well as unique and highly appreciated non-wood products (e.g., cork, pine kernels, mushrooms, medicinal plants, etc.).

Mediterranean forests have historically been subject to numerous threats (forest fires, over-exploitation, deforestation, soil degradation), today accentuated by climate change and alterations in land use patterns. In Portugal, human-induced ecosystem change started thousands of years ago, largely induced by the need to improve food production. This resulted in the decline of forests and changes in their composition. Therefore, by the end of the 19th century and beginning of the 20th, Portuguese forests were dominated by three main types: resinous stands, and managed native forests of cork oak (*Quercus suber)* and holm oak (*Quercus rotundifolia).* In this period, only around 15% of the national territory was covered by woodland and there were serious erosion problems in mountain areas. In order to mitigate these problems and increase forest coverage, the Portuguese government organised several forestation campaigns, mostly involving the planting of wild pine trees.

Since the beginning of the last century, important changes have occurred in woodlands, which have almost tripled in size to occupy around 40% of the Portuguese mainland, according to the most recent statistics. In fact, the woodland area increased considerably with the implementation of the 1938 afforestation plan on the remaining common land in the north and central regions of the country. In more recent decades, most afforestations have taken place on abandoned agricultural land as part of the set-aside strategy of the EU common agricultural policy. Historically the process focused primarily on exotic species, initially maritime pine (*Pinus pinaster*) and, inthe past 50 years, eucalyptus (*Eucalyptus globulus*), which proliferated due to intensive silviculture boosted by afforestation programs and the involvement of cellulose companies that rented or purchased land. However, parallel to the expansion of the woodland in recent decades, there has also been a generalized decline in their management, leading to increased stand-level fuel accumulation and landscape-scale fuel connectivity.

Even though fires have been a major feature of Mediterranean ecosystems for millennia, there has been a dramatic increase in the number of fires and amount of land burned in European Mediterranean areas in recent decades, principally due to climate change and alterations in land-use (Mouillot et al. 2002, Gonzalez and Pukkala 2007, Pausas and Fernandez-Munoz 2012). This trend, although encompassing some periods of lower burnt areas, distinguishes Portugal from other southern European States, as it has the highest number of ignitions and the greatest proportion of burnt areas, particularly in the central and northern regions. Forest fires therefore constitute one of the most relevant environmental problems and are frequently considered the major cause of loss of biological diversity.

Although, conservation and enhancement of biological diversity has become a critical issue of woodland management in Portugal over the past 25 years, other threats affect its sustainable development, including diseases (such as the pinewood nematode, *Bursaphelenchus xylophilus*), invasive plant species (particularly the *Acacia* species) and Mediterranean climate change (an increase in temperature and potential evapotranspiration, combined witha reduction in summer humidity).

This work analyzes the main causes of the changes that have occurred in the composition of Portuguese woodlands, mainly in the last century, and discusses the major threats to their sustainable management. In fact, there has been a massive reduction in the area occupied by native oak-dominated woodland (evergreen, deciduous and semi-deciduous) in Portugal, resulting from the combination of several factors over time. The remnant woods, usually restricted to exiguous patches in marginal areas, exhibit a structure and floristic composition shaped by management associated to human primary activities. Despite such negative effects, the dependence of traditional human activities on forest resources has sometimes supported the forests' own survival. But such protection ended with the abandonment of such activities.

Study Area

Mainland Portugal, which has a surface area of 89 015 km², is located in the Iberian Peninsula (Figure 1), in the extreme southwest of continental Europe. Portugal is bounded on the west and south by the Atlantic but has a Mediterranean climate, *Csb*, in the north and centre and *Csa* in the south, according to the Köppen climate classification. The average annual precipitation varies between around 1500 mm in the northeast and 500 mm in the southeast.

This pattern is reversed for the average annual temperature, with the highest figures registered in the Alentejo (Central and Lower Alentejo) and Algarve, and the lowest in the northernmost territories (Minho-Lima, Alto Trás-os-Montes).

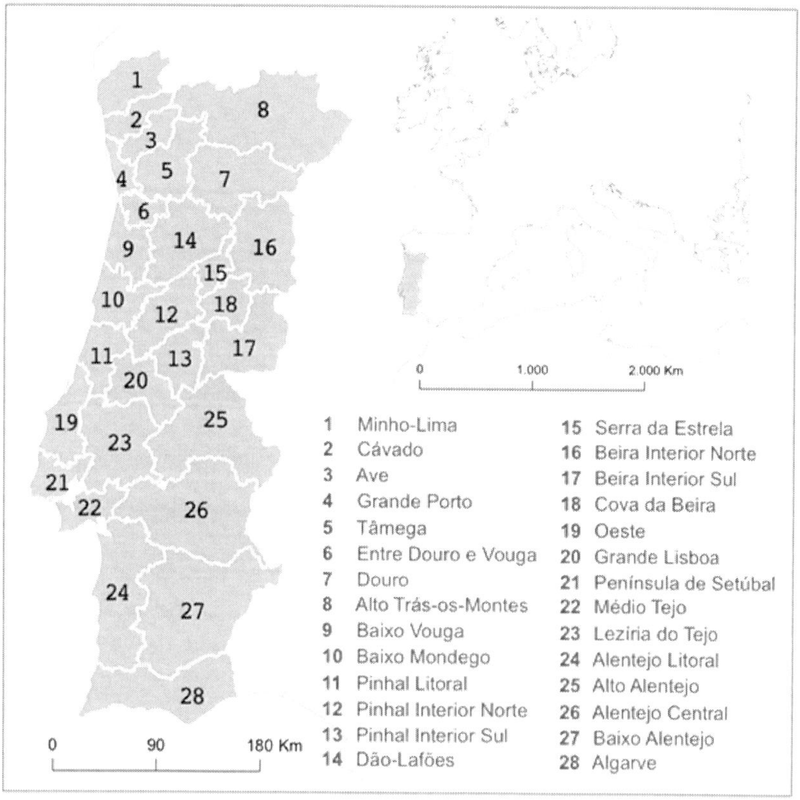

Figure 1. Sub-regions of mainland Portugal.

Despite this variability, the country as a whole reflects the seasonal pattern typical of a Mediterranean climate characterized by cool wet winters and hot dry summers. With the exception of the northwest region, the whole territory has a reasonably long dry season, lasting between three to five months, increasing from north to south and from coastal to inland Portugal.

In fact, the physical environment varies enormously between northern and southern Portugal. This variation is largely explained by the different physiographic characteristics; most of the country's mountains lie to the north of the River Tejo, where the landscape is more rugged, intersected by deep valleys. The highest mountain range is the Serra de Estrela, which runs northeast-southwest in central Portugal, reaching a maximum height of 1993 m. In these mountains, as in the northwest ranges, average annual precipitation exceeds 2000 mm.

The southern part of mainland Portugal has different physiographic characteristics. The Alentejo region (24, 25, 26, 27 – Figure 1) is mainly characterized by the vast flat or rolling terrain of the pediplain where the average altitude is approximately 200-300 m. The São Mamede mountain ridge, the highest in the Alentejo region with an altitude surpassing 1000 m, lies in the extreme north-east. In contrast, in the far south, the landscape is dominated by the two main Algarve mountain ranges, Monchique in the west and Caldeirão in the east.

Potential Distribution of Native Oak Forests in Portugal

With the exception of specific topographic positions (mountain summits) and edaphic conditions (alluvial plains, sand dunes), the whole territory of mainland Portugal would once have been covered by native forests dominated by the *Quercus* species (Silva, 2007). However, thesedeciduous and semi-deciduous forests are currently restricted to small and highly disturbed patches as a result of centuries of cutting, grazing, wildfires, and afforestation (activities that have left their fingerprints on the composition and structure of the remnant stands), reinforced by the use of maritime pine (*Pinus pinaster*) and exotic species (*Eucaliptus* sp., *Cupressus lusitanica*, *Pseudotsuga menziesii*)..

This trend towards the reduction in the area of oak forests was still evident in the last century. The implementation of directives designed to increase cereal production in mainland territory and the massive expansion of traditional activities in rural areas (agriculture, grazing), as a result of the significant population increase in the 50s and 60s, contributed decisively to restricting the native oak forests to small enclaves, particularly in n the territory north of the River Tejo.

Although this trend subsequently went into reverse, with the massive abandonment of traditional activities in the last four decades, the oak forests have had no chance for recovering from such deep and perniciouseffects suffered over centuries. Despite the fact that there has been an increase in forested areas, this tendency is based on afforestation processes using faster-growing species of faster growth (e.g. *Pinus pinaster* Aiton), some of which are exotic (e.g. *Eucaliptus globulus* Labill., *Quercus rubra* L.). So, alongside the loss of area, there have also been significant changes to theforest composition, structure and biodiversity. Such changes (particularly the use of exotic species for afforestation) are interpreted as both the cause and consequence of abandonment, as it hashelped breakthe relationship between the traditional activities and the native forest, reinforcing the trend towards the depopulation ofrural areas (Paiva, 2007a, 2007b).

Despite being potentially controversial, the abandonment of traditional activities in rural areas is doubly negative for the native oak forests. Although the residual patches experienced continuous disturbance due to the cyclical collection of understory biomass and leaf litter, which contributed to changes in their floristic composition, some were conserved for precisely this purpose. After the abandonment of agriculture, such woods, some of which present an open structure, registered an increase in understory biomass, alongside the increasing dominance of shrubby communities in the landscape. These conditions created a scenario of biomass contiguity, a scenario that has become deleterious in wildfire contexts.

Despite such negative effects, the dependence of traditional human activities on forest resources has supported their survival, very often reducing the impacts ofwildfire. However, this balance has been broken with the abandonment of such activities.

The description of the potential forests is supported on a phytosociological perspective (Géhu and Rivas-Martínez, 1981). The taxonomic nomenclature in use follows Castroviejo et al. (1986-2012) and Franco and Afonso (1998). The biogeographic nomenclature and the proposal for the potential distribution of the oak forests are based on Costa *et al.* (1998) (Figure 2). The bioclimatic classification follows Rivas-Martinez (2004).

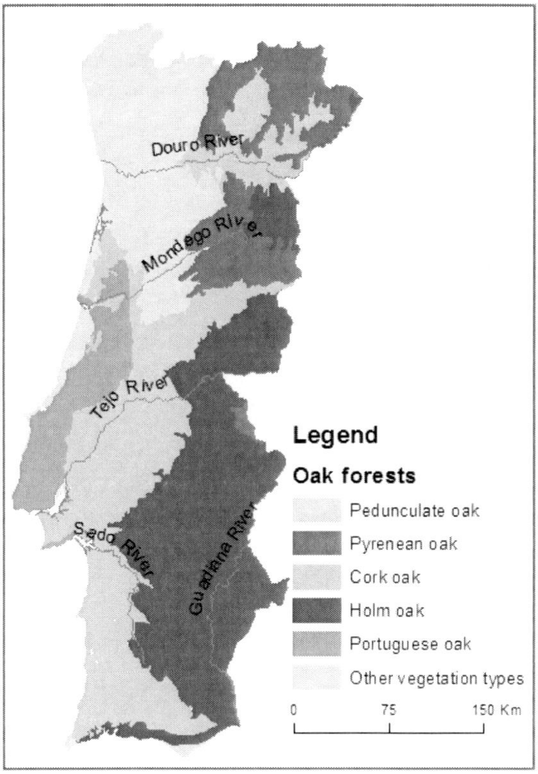

Figure 2. Potential distribution of native oak forests in mainland Portugal.

DECIDUOUS AND SEMI-DECIDUOUS OAK FORESTS

Pedunculate Oak Forests *(Quercus robur* L.)

The pedunculate oak forests are found mostly in the western territories of northern and central mainland Portugal (Figure 2), which is their southernmost limit of distribution in the Iberian Peninsula. In these areas, the climate is strongly influenced by the ocean, as can be seen by the higher levels of annual precipitation and fewer dry months, configuring a context of buffered climatic conditions in mainland Portugal. As this species requires a great deal of water, such forests are found in areas with low water-deficit (< 2 dry months = P (mm) < 2*T (ºC)), high precipitation (1000 – 2000 mm) and narrow annual temperature ranges. In these territories, the forests potentially occupy low and middle altitudes (up to 800-1000m), as they are not very tolerant of low temperatures. Above such altitudes it is replaced by the Pyrenean oak (*Quercus pyrenaica* Willd.), as can be seen on higher peaks in ranges such as Cabreira, Peneda, Amarela and Gerês (Carvalho et al., 2007b).

Most of the potential territory is associated to one version of forest, *Rusco aculeati-Quercetum roboris*, which occupies the lower altitudes. Middle elevations, whereprecipitation is higher and average temperatures lower, are associated to a more temperate version (*Vaccinio myrtilli – Quercetum roboris*) (Honrado, 2003).

Though dominated by the pedunculate oak, these forests frequently include other temperate trees, such as*Acer pseudoplatanus, Ilex aquifolium, Sorbus aucuparia, Betula celtiberica, Corylus avellana* andsome termophylous *taxa* (*Prunus lusitanica, Laurus nobilis, Viburnum tinus, Ruscus aculeatus, Tamus communis*), which benefit from the buffered climate. At the upper areas of distribution, in supratemperate areas, there is an understory floristically associated to *taxa* with Eurosiberian distribution (*Vaccinio myrtilli – Quercetum roboris*). At higher altitudes and further inland, where the annual temperature ranges are greater, such forests are replaced by the Pyrenean oak forests (*Quercus pyrenaica*).

The main factors contributing to the restricted extent of pedunculate oak forests nowadays are agriculture, grazing activities and wildfires (Carvalho et al., 2007b). The high quality of its wood, which makes it suitable for a wide range of applications (e.g. furniture, house construction, shipbuilding, cooperage), has also contributed to the annihilation of such forests (Paiva, 2007b). According to the last National Forest Inventory (2005-2006), pedunculate oak forests (mature stands) covered an area of approximately 37000 ha (Carvalho et al., 2007b).

Pyrenean Oak Forests (*Quercus pyrenaica*)

Pyrenean oak forests are associated to sub-humid and humid mesomediterranean to supramediterranean bioclimatic territories of northern and central inland Portugal (Figure 2). Considered submediterranean in terms of distribution, they occupy transitional areas between temperate climate zones, strongly influenced by the Atlantic Ocean tothe west, and Mediterranean zonesat lower altitudes and to the south. Thus, it forms atransitional species between the Atlantic temperate deciduous forests dominated by the pedunculate oak (*Quercus robur*) and the Mediterranean evergreen forests of cork oak (*Quercus suber*) and holm oak (*Quercus rotundifolia*).

Theseforests are found mostly at medium to high altitudes (700 – 1200m) within the easternmost sections of the Douro and Mondego catchment areas, coinciding withthe *Carpetano-Iberico-Leonese* biogeographic province (Costa et al., 1998). However, theyalso occur on the higher elevations of the *Galaico-Portuguese* province in the NW (a territory associated with forests of *Quercus robur*) and on middle and high elevations in the easternmost territory of the Tejo catchment area. Most of the territories of potential distribution are siliceous in nature (schists and granites).

According to the last National Forest Inventory (2005-2006), Pyrenean oak forests currently exhibit the largest occupancy area of all deciduous and semi-deciduous oak forests (approx. 80 000 ha) (Capelo and Catry, 2007a), yet theirtheir potential distribution area is the smallest.

The Pyrenean oak forests are highly diverse with regard to theirfloristic composition and very heterogeneous. In terms of climate, astheir distribution area is associated to transition conditions, theycan be divided into three types in accordance with the floristic composition of the understory (, a fact that is related to the extent of the seasonalwater deficit) (Carvalho et al., 2007a). Phytosociologically, two widely distributed vegetation series dominated by *Quercus pyrenaica*[1] have been identified in climatophyllous positions: a northern more humid version (*Holco mollis-Quercetum pyrenaicae*), which includes more *taxa* with Atlantic affinities, and a southern one (*Arbuto unedonis-Quercetum pyrenaicae*), which reveals theinfluence of Mediterranean climatic conditions. Its presence at such low latitudes is associated to small mountain ranges (S. Mamede), surrounded by predominantlyevergreen forests. Some authors (Costa et al., 1998) claim that this constitutes another series (*Genisto falcatae-quercetum pyrenaicae*); however, the fact that it shares a large amount of *taxa* with the first suggests that they aresynonymous (Figueiredo, 2005).

Despite the general paucispecific conditions at the tree layer, the more temperate version can include wild cherry (*Prunus avium*), pedunculate oak (*Quercus robur*), common holly (*Ilex aquifolium* L.), rowan (*Sorbus sp.*) andbirch (*Betula sp.*) in supramediterranean territories. In mesomediterranean areas, the greateravailability of soil moisture can even support a forest co-

[1] We assume that the concepts *Holco-quercetum pyrenaicae* and *Genisto falcatae-quercetum pyrenaicae* are synonyms, once floristic composition are very similar.

dominated with *Fraxinus angustifolia* (*Fraxino angustifoliae-Quercetum pyrenaicae*).

The more Mediterranean version is characterized by the significant presence of the strawberry tree (*Arbutus unedo* L.), and can include several tree *taxa* associated to the neighboring woods of cork oak, holm oak and Portuguese oak, sharing with them several species of the understory layer.

Although the wealth of current species in the understory is dependent on natural aspects, such as the density of tree cover, and edaphic and prevalent climate conditions, it also seems to be related to the length and type of disturbance that the forests have beenexposed to. In fact, their structure, composition and succession are severely affected by the length, intensity and type of use to which they are subjected. It is generally agreed that higher degrees of disturbance are associated to reduced plant wealth (Carvalho et al., 2007a). Historical land-use has not only dictated a massive reduction in the area of such forests but has also shaped their floristic composition, very often associated to a significant impoverishment of such communities due to the frequent removal of the understory biomass and litter. Changes in floristic composition at tree level also occur as a result ofmanagement, and usually involve a preference for spontaneous faster-growing species such as *Pinus pinaster* Aiton (often a source of propagules for neighboring afforested areas).

As with other native species, these forests are currently reduced to small patches in a landscape dominated by shrubby communities.

Portuguese Oak Forests (*Quercus faginea* Lam.)

The Portuguese oak is a semi-deciduous species with a prevalent Ibero-African distribution. It is found particularly in the SW Iberian Peninsula and NW Africa, with a few stands in SW France and the Balearic Islands (Franco, 1990; Paiva, 2007a).

Woods dominated by thisoak frequently include other oak species, such as cork oak (*Quercus suber*), Pyrenean oak (*Quercus pyrenaica*), pedunculate oak (*Quercus robur*) and holm oak (*Quercus ilex*), especially at the edges of its potential distribution. The understory has significant plant wealth, including several Mediterranean evergreens (*Viburnum tinus*, *Rhamnus glandulosa, Myrtus communis*), prickly shrubs (*Crataegus monogyna*, *Ruscus aculeatus*), liana-habitat species (*Lonicera peryclimenum*, *Tammus communis*, *Rubia peregrina*) and herbaceous *taxa* (*Paeonia broteroi*, *Hyacinthoides hispanica*, *Gennaria diphylla*).

Most Portuguese oak stands are dominated by the taxon *Quercus faginea* subsp. *broteroi* (Cout.) *A. Camus*, the only sub-species that supports the definition of pure stands and which has the largest area of potential distribution. The forests of *Quercus broteroi* occur in thelimestone territories in the west of the central region of mainland Portugal and in the southern mountain ranges of Serra Algarvia (southern Portugal), where its distribution is very restricted. In the latterlocation, the Portuguese oak forests may also include *Quercus faginea* subsp. *alpestris* (a sub-species mostly restricted to NW Africa) in submountain areas (*barrocal*) (Capelo and Catry, 2007a). The other recognized sub-species, *Quercus faginea* subsp *faginea* is mostly restricted to mesomediterranean sections of the tributary valleys of the River Douroin NE Portugal, and is associated to siliceous substrates. In this case the definition of pure stands is very rare, and generally occurs after the disturbance of forests usuallydominated by the cork oak, where the Portuguese oak can at most achieve co-dominance. Its occurrence in mixed forests is determined by the narrow distribution of the *taxon* ata marked altitudinal gradient in a topographic context dominated by valleys, with cork oak forests at lower altitudes and Pyrenean oak forests at higher ones.

Quercus broteroi's tolerance of limestone soils and precipitation of between 500 and 1000 mm can explain its potential dominance in the western part of the central region. In sandstone lithology, Portuguese oak forests are replaced by cork oak forests (*Asparago aphylly-Quercetum suberis*).

As with other forests, thosedominated by this subspecies are currently restricted to small patches. The best preserved stands can be found in the limestone substrates in the western sector of mainland Portugal, particularly on medium-altitudemountain ranges, such as Estremenho Limestone Massif; Arrábida, Sintra, Montejunto; Sicó, Alvaiázere (Capelo and Catry, 2007a).

In the western part of the central region, the woods dominated by the Portuguese oak (*Arisaro- Quercetum broteroi*) suffered a significantloss of area between 1972 and 1995. This amounted to around 50% in some territories, but was as high as 90% in others (such as the municipality of Montemor-o-Velho) (Bingre and Damasceno, 2007).

In limestone areas, the scarcity of deep soils suitable for agriculture has definitively contributed to the annihilation of thisforest. In addition, as the wood is of little value for woodwork (Fabião et al., 2007), it has often been replaced by other species (holm oak, cork oak, pine), wherever the lithology permits (Bingre and Damasceno, 2007).

It has a good rate of natural regeneration, better from resprout than seedlings (Fabião et al., 2007). Thus, these woods can be recovered without intense intervention.

EVERGREEN OAK FORESTS

Evergreen oak forests are dominant in the southern half of mainland Portugal (a territory characterized by long and dry summers), where a specific type of management, the *montado*, has ensuredthe preservation of a modified version of the native woodlands. Traditionally, acombination of activities (forestry, extensive agriculture and cattle production) supported the maintenance of open stands, created through the clearing of the native forest. Covering large areas, thislow-intensity system is implemented after reduction of the area occupied byforest and a thinning of the tree layer in areas reserved for the exploitation of forest resources and grazing. Cork productivity actually increases with thisopening-up of the forest (Costa and Pereira, 2007).

In fact, it was low agricultural productivitythat prompted the quest for supplementary sources of profit through the exploitation of forest resources. Unfortunately, the equilibrium achieved has now been upsetby modernization, not only the introduction of heavy machinery but also changes in the type of agriculture practised (that is to say, a switch to intensive versions of agriculture driven by an increase in the availability of water through dam construction)has led to)..

The northern limit of the evergreen oak forests is marked by a lineof mountains in the centre of mainland Portugal (Cordilheira Central, Serras d'Aire and Candeeiros, Serra de Montejunto, Serra de Sintra) (Figure 2).

Holm Oak Forests (*Quercus rotundifolia* Lam.)

Holm oak forests are dominant in the driest areas of the mainland territory, i.e. from dry to sub-humid thermo- and mesomediterranean conditions. They occur in areas where annual precipitation is under 500 mm and the annual temperature range is significant. In the driest mesomediterranean areas, found in southern inland territories in the lowest section of the Guadiana valley, there is a predominance of termophylous trees including *Myrtus communis* (*Myrto communis-Quercetum rotundifoliae*). However, most of the territory is occupied by a mesomediterranean sub-humid species, *Pyro bourgeanae-Quercetum rotundifoliae*, replaced by cork oak in areas with higher precipitation.

Over the centuries, thesewoods were managed in association withthe *montado* system. As a complex system of resource exploitation, this extensive landuse caused significant disturbance tothe native woods, bringing changes to their structure and evident floristic impoverishment.

Cork Oak Forests (*Quercus suber* L.)

Distributed throughoutthe western Mediterranean region inboth Europe (Portugal, Spain, Italy, France) and Africa (Morocco, Algeria, Tunisia), cork oak forests are particularly widespread in Portugal, with an area of some 736,700 ha (2006). This is approximately 40% of their total area (Mendes, 2007).

In Portugal, cork oak forests are clearly associated to areas with a Mediterranean climate pattern (i.e. long dry summers, and low precipitation concentrated into the period OctoberMay). They occupy areas where there is little variation in the annual temperature and a range of precipitation of 500 - 800 mm (Costa and Pereira, 2007). Because of their sensitivity to low temperatures, cork oak forests are dominant in areas where there is more oceanic influence (i.e. in the southern half of mainland Portugal and lower territories of the Douro basin). In territories with precipitation below 500mm and greater annual temperature ranges, the cork oak forests are replaced by holm oak forests (Capelo and Catry, 2007b), while in areas with lower temperature, usually at higher altitudes (>700m), they are replaced by forests dominated by the Pyrenean oak.

Theseforests have suffered great disturbance due to centuries of management and the occurrence ofwildfires, which have increased in frequency in the last four decades. The vast majority (aprox. 70%) has undergone different levels of exploitation under the *montado* system (Costa and Pereira, 2007), to the extent that low-disturbed fragments are very rare.

Despite the economic value of the cork oak, its area has decreased in the last five decades (Capelo and Catry, 2007b). Recurrent wildfires, the low woodland recovery rate and the introduction of heavy machinery have made contributed to thisprocess.

The northernmost area of potential cork oak distribution is the interior part of the Douro valley and its tributaries, particularly the northern bank. These forests occupy dry to humid mesomediterranean conditions. Although the potential area is considerable, such woods (*Rusco aculeati – Quercetum suberis*) are currently very restricted as a result of extensive use of the territory for agriculture (e.g. wine production, olive oil). This level of disturbance is common to all territories potentially associated to cork oak forests. Nevertheless, cork oak forests do have a significant presence south of the River Tejo, particularly in the western areas. In fact, it is possible to find extensive areas of (usually managed) open forest, a scenario that has no similarity in terms of forests dominated by native oaks. Despite their progressive decrease, the persistence of open forests is related to the exploitation of cork.

Given its considerable soil moisture requirements and preference forlong summers, it predominates in the westernmost areas of the southern territory. In upper mesomediterranean sub-humid to humid territories, cork oak forests are replaced by semideciduous Pyrenean oak forests, namely the *Arbuto unedonis-Quercetum pyrenaicae* in the Upper Alentejo and the *Genisto falcatae-Quercetum pyrenaicae* in Alto Trás-os-Montes. In dry to sub-humid thermomediterranean and dry mesomediterranean conditions, cork oak forests are replaced by holm oaks.

Most of the southern part of its potential territory has thermomediterranean humid to mesomediterranean sub-humid conditions. The thermomediterranean areas are associated to a more thermophylous version of forest (*Myrto-Quercetum suberis*), while the mesomediterranean territories tend to have a more humid version (*Asparago aphylli-Quercetum suberis*). The southernmost territories, in the Algarve, are associated to a more open forest, which includes the wild olive (*Oleo-Quercetum suberis*). Here, the cork oak forest is replaced by communities dominated by the holm oak (*Smilaco-Quercetum rotundifoliae*) and by the Portuguese oak (*Arisaro-Quercetum broteroi*) (Costa et al., 1998).

The Development of Woodlands in Mainland Portugal: The Last Century

Mainland Portugal

Mediterranean ecosystems have historically been subject to a long-term human influence with both positive and negative effects (Hobbs et al. 1995). Consequently, the ecosystems of the Mediterranean basin have been defined as "total human ecosystems" (Naveh 1987). In Portugal, the human-induced change in ecosystems also started centuries ago. The need to improve food production for a growing population led to the decline of forests. Although this was a continuousprocess, it is possible to identify periods when there was a significant reduction in the oak forests in association with specific socio-economic dynamics. This trend can be deduced from pollen sequences, which show a reinforcement in *taxa* related to shrubby communities (Castro et al., 2005; Capelo and Catry, 2007; Carvalho and Devy-Vareta, 2007).

The last massive reduction in the pollen sequences, dated to 800 years ago, is associated to an increase in cereal production, coinciding with Portugal's acquisition of political autonomy. The subsequentdevelopment of shipbuilding and shipping industries (between the 15th and 18th centuries) gave another important impulse to the extermination of the oak forests.

Thus, by the end of the 19th century and beginning of the 20th, around 10-15% of the national territory was covered by woodland, divided into three main categories according to species composition: i) the resinous category, associated to afforestation processes based on the use of pines (*Pinus pinaster, Pinus sylvestris)*; ii) the *Quercus suber* and *Quercus rotundifolia* category; iii) and (on a smaller scale) the other *Quercus* and *Castanea* species. So at the beginning of the century, Portugal would have been a deforested country, and the woodland would have had a different composition from the natural forest described earlier. Serious erosion problems in mountain areas were already being mentioned.

In fact, the Portuguese mainland has a very long history of deforestation (Devy-Vareta, 1985, 1986) as well as of efforts to protect woodland, through state intervention. According to Caldas (1978), such protection efforts date from the first dynasty (13th and 14th centuries) and were expressed by the demarcation of "coutadas" belonging to the king and nobility in order to preserve some spaces from the action of humans and livestock.

Later, in 1494, a law was promulgated to compel the planting of a number of trees (Devy-Vareta, 1986), while the 1565 "tree law" decreed the planting of pines (or chestnut, oak and other trees adapted to the soils) on municipal wastelands. However, in 1888, the continued depletion of plant cover and increasing concern about soil erosion led to the creation of forestry administrations in the mountainous areas of Gerês and Estrela with the aim of reforesting 90,000 hectares (Rego, 2001). Since then, the Portuguese government has promoted several campaigns for forestation, mostly involving common land and the planting of maritime pine trees.

With the implementation of the 1938 afforestation plan on the remaining common land of the northern and central regions, the forested area increased considerably. In fact, 30 to 50% of commons and 13 to 22 % of the total area of these municipalities were reforested (DGSFA 1953). The favoured species was *Pinus pinaster* Aiton (maritime pine) (Baptista, 1993; Coelho, 2003, DGDR/AFN, 2006), due to its fast growth and the lowcost of seeding. Consequently, this became the most prevalent species in the national forest. According to Baptista (1993), 1,419,000 ha were reforestedbetween 1931 and 1973, and as a result, its geographical expression almost tripled in the first half of the 20th century from 430,000 ha to 1,234,000 ha (Figure 3).

In the 1965/74 National Inventory, the area covered by the *Pinus pinaster* peaked at around 1,370,000 ha. After that, there was a clear decrease, with an area of 885,000 ha registered in the last forest inventory (2005/06).

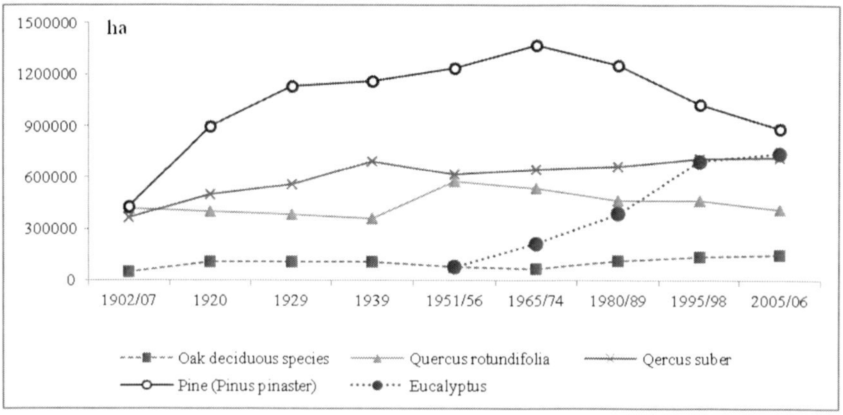

Source: 1902/07- Evaluation based on the Agricultural Charter (*Carta Agrícola)* of 1902/1907; Sertório de Monte Pereira, "A produção Agrícola", 1908; Agricultural Statistics, 1959, INE (National Statistics Institute); 1920 and 1939: Estimation by Henrique de Barros, M. Azevedo Gomes and E. Castro Caldas in *Evolução da Agricultura Portuguesa*, 1945; Estimation of J. Rasteiro and Mendes de Almeida in *Portugal - Agricultura e Silvicultura*, 1929; 1951-56: Service of Agrarian Recognition and Management, Agricultural statistics, 1965, INE; 1965/74: National Forest Inventory,. Agricultural statistics, 1980, INE; 1995/98: 4th National Forest Inventory (IFN4), DGRF (National Forestry Services), *Anuário Florestal*, 2002; 2005/06: 5th National Forest Inventory (IFN5), DGRF.

Figure 3. Development of the main species in Portuguese forests (20th to21st centuries).

Although this tree still dominates the Portuguese woodland, another species, the eucalyptus, has undergone great expansion. Indeed, its area has increased from less than 100,000 ha to 740,000 ha since the second half of the last century (an increase of 800%) (Figure 3).

The *Eucalyptus globulus* Labill., an exotic species which was introduced in the 1950s, is widespread all over the country. Due to its rapid growth, it is much in demand for the paper pulp industry, which is one of the main reasons for its rapid expansion (MEA, 2004).

Nowadays, almost 3 350 000 ha (40%) of Portuguese national territory is covered by forest. The main species are pine, eucalyptus, *Quercus suber*, *Quercus rotundifolia*, and to a lesser extent the other deciduous oak species. Although cork oak (*Q. suber*) has a very restricted distribution in the world, its largest continuous area is located in the Iberian Peninsula, especially in the southwest (Silva and Catry 2006). In Portugal, cork oak is found especially in the south, where it forms mono-specific stands, but also in the rest of the country, often mixed with other species.

Natividade (1950) distinguishes typical forest stands (that is, denser stands which are mainly used for the production of cork) from those which are part of the agroforestry system called the *montado*. This system (also known as *dehesa* in Spain) comprises scattered *Quercus suber/Quercus rotundifolia* (evergreen oak) trees sharing the same space with agriculture or pasture. This mixture has many advantages, as the trees are a source of additional income for the landowner (cork, firewood and fodder provided by foliage and acorns) and guarantee soil protection. Other land uses that have been gaining importance in the *montado* areas are hunting and the breeding of a regional race of pigs. This latter activity was, in fact, the main historical reason for the expansion of the *montado* system in the 17th century (Coelho, 1996). The growing demand for pork meat, as well as cork, led to an increase in the "montados" (cork-trees and holm-oaks) of the Alentejo (MEA, 2004). By the middle of the 20th century, the forest area had doubled (Figure 3), with cork oak stands increasing from 366,000ha to 716,000 ha and the holm-oaks from 417,000 ha to 573,000 ha.

However, since the second half of the last century, the area covered by holm-oak stands has been decreasing. This could be ascribed to the abandonment of traditional farming and consequent loss of economic interest in this species. Conversely, the opposite trend observed in the cork oak stands could be related to the higher economic interest of this species. Another reason (probably the main one) for the conservation of this system in large areas of Portugal is the existence of very ancient (at least since the 13th century, Mendes, 2002) legislation protecting cork oak stands.

Deciduous and semi-deciduous oak species (the native forest of the largest areas of northern and central regions) occupya very limited area: less than 50,000 ha at the beginning of the 20th century, and 150,000 ha at the beginning of the 21st. Although insignificant in the total woodland (it accounts for around 5% of the total forest cover), the deciduous oak forest has tripled its geographical area over the last century. This increase in the autochthonous vegetation is related to the sequential abandonment of traditional land use throughout the second half of the 20th century, coinciding with widespread migration to urban centres and to certain European countries (France, Germany, Luxembourg, etc.) (Nunes, 2007, Almeida et al., 2009). This exodus and consequent abandonment of cultivated land has resulted in enhanced natural secondary succession and the spread of shrub and woodland (Nunes et al., 2010).

According to Godinho-Ferreira et al. (2005), despite the increase in forest area and the reduction in the disturbance generated by traditional activities in rural areas, the native oak forests have maintained the same spatial restriction and disturbed condition in a scenario massively dominated by shrubby communities. Several other factors have contributed to the reduction in the importance and occupancy of native oak forests: failed conservation attempts; the prevalence of policies focusing on fast-growing species; and unfitting directions on forest planning and management attending the potential conditions of the territory.

On the other hand, in recent decades, the EU Common Agrarian Policy (CAP) has also encouraged this move away from cultivated land, in particular for cereal crops. In order to compensate farmers for the reduction of direct support, follow-up measures have beenimplemented, including agri-environmental and afforestation measures. More than 900,000 ha of forest land benefited from CAP measures from 1986 to 2006, 217,000 ha of which corresponded to the afforestation on abandoned agricultural land (DGRF/AFN, 2006). In fact, pine and especially eucalyptus forest thereafter became alternative income sources derived from the abandoned agricultural land. Both factors have been responsible for the increase in the woodland area in Portugal since the second half of the 20th century.

Regional Distribution

Figures 4, 5 and 6 show the regional distribution of various species on the basis of National Forestry Inventories carried out in the 60s and 70s and more recently, in 2005/06. As regards the incidence of *Pinus pinaster*, the densest areas are in the centre of Portugal in the regions of Dão-Lafões, Pinhal Interior Norte, Pinhal Interior Sul and Pinhal Litoral (Figure 4) (in fact, the preponderance of this species has determined the toponymy of these regions, as "pinhal" means "pine wood"). However, more recently there has been a significant decrease in the spatial significance of this species, which is the one most seriously affected by forest fires.

Exceptions to this trend include Altos Trás-os-Montes, where the area covered by pine has almost doubled, and Beira Interior Sul, Alentejo Litoral and Algarve, where there have been more modest increases.

This increase in northeastern Portugal could be associated to the implementation of agricultural land afforestation measures (Regulation 2080/92) under the Common Agricultural Policy. This provided support for the afforestation of former agricultural lands, aiming also to enhance long-term forest resources, and combat soil erosion and desertification by promoting forestry as an alternative form of land use. However, the greatest increase was recorded in the eucalyptus area, which tripled its expression in mainland Portugal, making it the second-most important species in the country (Figure 4).

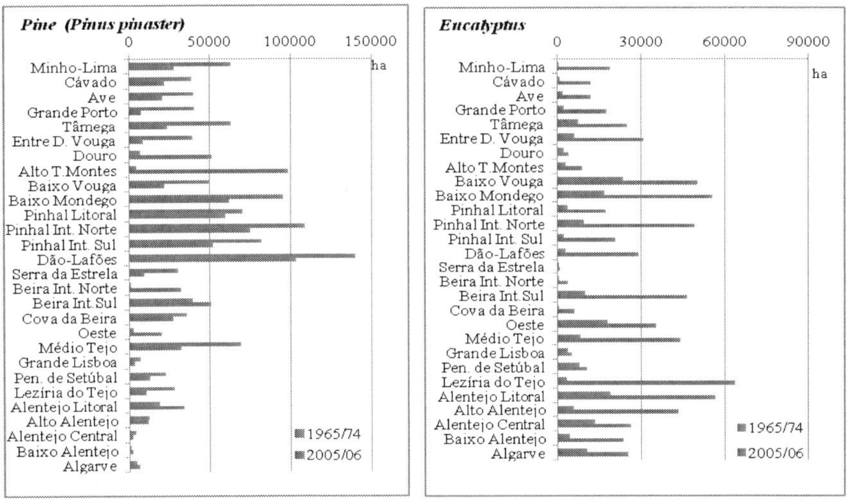

Source: 1965/74- National Forest Inventory, by main species. Agricultural statistics, 1980, Statistics Portugal. 2005/06- 5th National Forest Inventory (IFN5), General Direction of Forestry.

Figure 4. Regional variation in Pine and Eucalyptus area, between 1965/74 and 2005/06.

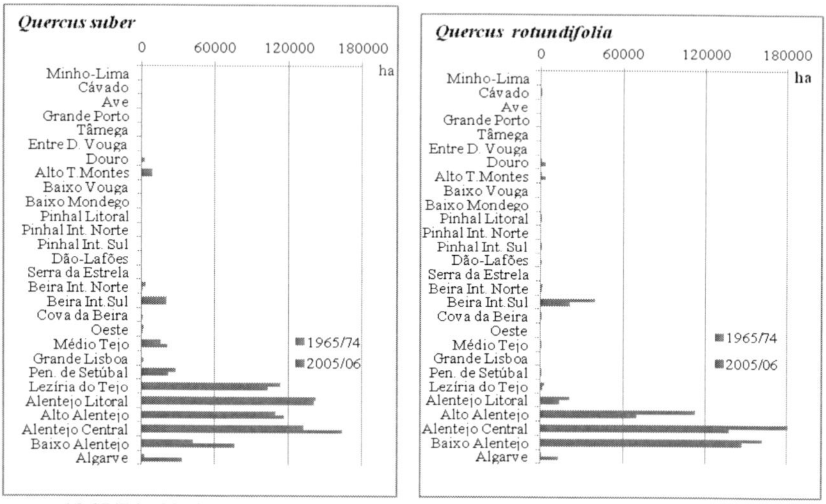

Source: 1965/74- National Forest Inventory, by main species. Agricultural statistics, 1980, Statistics Portugal. 2005/06- 5th National Forest Inventory (IFN5), General Direction of Forestry.

Figure 5. Regional variation in *Quercus suber* and *Quercus rotundifolia* area, between 1965/74 and 2005/06.

All regions have registered an increase since the ‘60s and ‘70s, most pronounced in northwestern and central regions. *E. globulus* has assumed an important role not only in terms of area, but also economically, providing the raw material for the pulp and paper industry, one of the most important industries in the country. Eucalyptus plantations have emerged as an alternative to other land uses, competing with grasslands, rain-fed agriculture, shrublands, maritime pine forest, and, more recently, as an alternative to land abandonment (Soares et al., 2007). South of the River Tejo, several areas were afforested with *Pinus pinaster* and, more recently, with *Eucalyptus globulus* Labill., although the forest was dominated by perennial oaks, such as the cork oak (*Quercus suber*) and holm oak (*Quercus rotundifolia*). These two species compete mainly with cork oak, requiring similar soil and climate conditions.

Although more than 70% of the cork oak stands are situated in the Alentejo region, this species would once have occupied a much larger area of Portuguese territory, particularly in the northwest and central west regions, where it is almost absent nowadays (Silva and Catry, 2006). Nevertheless, Figure 5 shows that there are still considerable areas of cork oak south of the River Tejo, while in some regions, such as Central and Lower Alentejo, these are increasing (though elsewhere, as with the Lezíria do Tejo and the Setúbal peninsula, the tendency has been in the opposite direction). While the cork oak's main area of distribution is in the Alentejo, the holm oak's natural centre lies further eastwards. The holm oak holds fourth place in terms of land cover, though it is decreasing, if we compare the results obtained in the 1965/74 inventory with those of the last one in 2005/06, when it still occupied around 413, 000 ha.

This means a loss of 23% in three decades. The reasons for this drastic decrease include a loss of economic interest in this species, as well as the general abandonment of traditional farming in rural areas of Portugal. Another factor is increasing oak mortality.

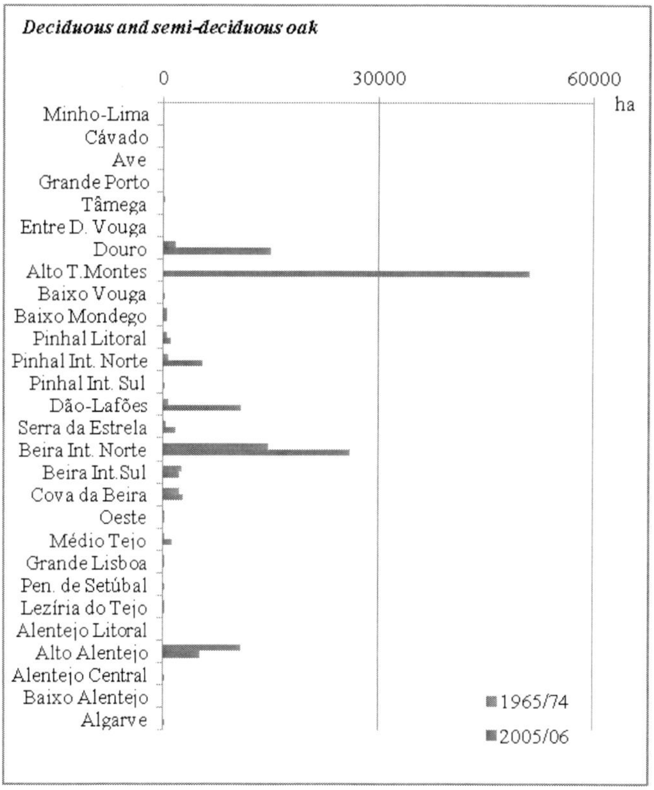

Source: 1965/74- National Forest Inventory, by main species. Agricultural statistics, 1980, Statistics Portugal. 2005/06- 5th National Forest Inventory (IFN5), General Direction of Forestry.

Figure 6. Regional variation in *Deciduous* and *semi-deciduous* area, between 1965/74 and 2005/06.

In fact, in the National Forest Strategy, the Forest Directory (DGRF, 2006) noted an ongoing process of disease in the two oak species *Q. suber* and *Q. rotundifolia* since the 1970s, particularly since 1997.

Indeed, there has been a massive reduction in the area of occupancy of native forests dominated by oaks (deciduous and semi-deciduous), resulting from the combination of several factors over time (Figure 5). The abandonment or marginalisation and collapse of traditional farming systems in some regions, such as Beira Interior Norte, Dão-Lafões and Alto Trás-os-Montes, explain the recovering with *Quercus,* mainly, *Quercus pyrenaica.* Nevertheless, this native tree covers negligible areas, because restoring native vegetation is a very long process, disturbed by the regular occurrence of forest fires.

Threats to the Sustainable Management of Woodlands

Forest Fires

Fire is an ancient and universal ecological force which has shaped most of the plant communities and landscapes of the Mediterranean Basin. In Mediterranean ecosystems, fire is both a significant ecological disturbance and a regenerative process, although the fire regime has changed in recent decades. Fires have increased in intensity and extent, losing their beneficial ecological role and becoming catastrophic events (Noss et al., 2006). In the Portuguese mainland over the last 3 decades, the number of forest fires has exceeded half a million ignitions and the total burnt area was approximately 3 582 800 ha, representing more than a third of the surface area of mainland Portugal (Nunes, 2012).

Figure 7 shows the total burnt area classified as wood and shrubland, for the period under analysis. Around 52% of the burnt area was classified as shrubland whilst 48% was woodland.

Pine and eucalyptus forests are highly flammable, which leads to large and severe fires (Doerr et al. 2000). In the last 30 years, 2003 was the worst fire season in Portugal, resulting in the burning of almost 430 000 hectares of woodland (around 286 000 ha), shrubland and crops, with overall economic losses amounting to 1 200 million euros (DGRF, 2006).

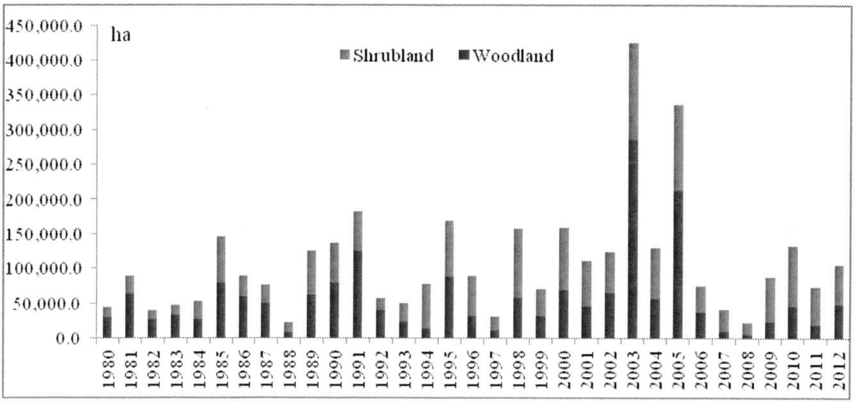

Source: Portuguese Government Forestry Department.

Figure 7. Evolution of burnt area, classified as wood and shrubland, over the period 1980-2012.

Twenty-one human lives were lost, hundreds of homes and agricultural premises destroyed and thousands of farm animals killed. As many as 325 000 hectares were burnt in 2005, of which 230 000 ha was woodland.

Historically, the centre (Beira Interior, Serra da Estrela, Pinhal Interior Norte and Sul) and the northern regions (mainly Minho-Lima, Tamega and Alto Trás-os-Montes) of mainland Portugal have been the areas most seriously affected by forest fires. These regions account for more than 50% of the total area burned within the 33-year period.

Southern Portugal has the lowest totals for burnt areas during the period under review (Figure 8). In fact, one of the characteristics of the *montado* system is its low vulnerability to fire (Silva and Catry, 2006). Understory removal for crop or pasture cultivation is responsible for a very low fuel load, which considerably reduces fire hazard. This explains the low fire 'preference' for *montado*s (Nunes et al., 2005) when compared to other forest systems in Portugal.

Although climatic and meteorological parameters are key elements in the occurrence of fires in Portugal, these factors alone do not justify the very large burnt area and the dangerous situations faced by many people living in rural areas and wildland-urban interfaces. In fact, many studies suggest that the increasing incidence and impact of forest fires in Mediterranean environments can be mainly attributed to the decline of the landscape mosaic that has historically characterized Mediterranean rural areas (Leite et al., 2011; Moreira et al., 2011; Nunes, 2012).

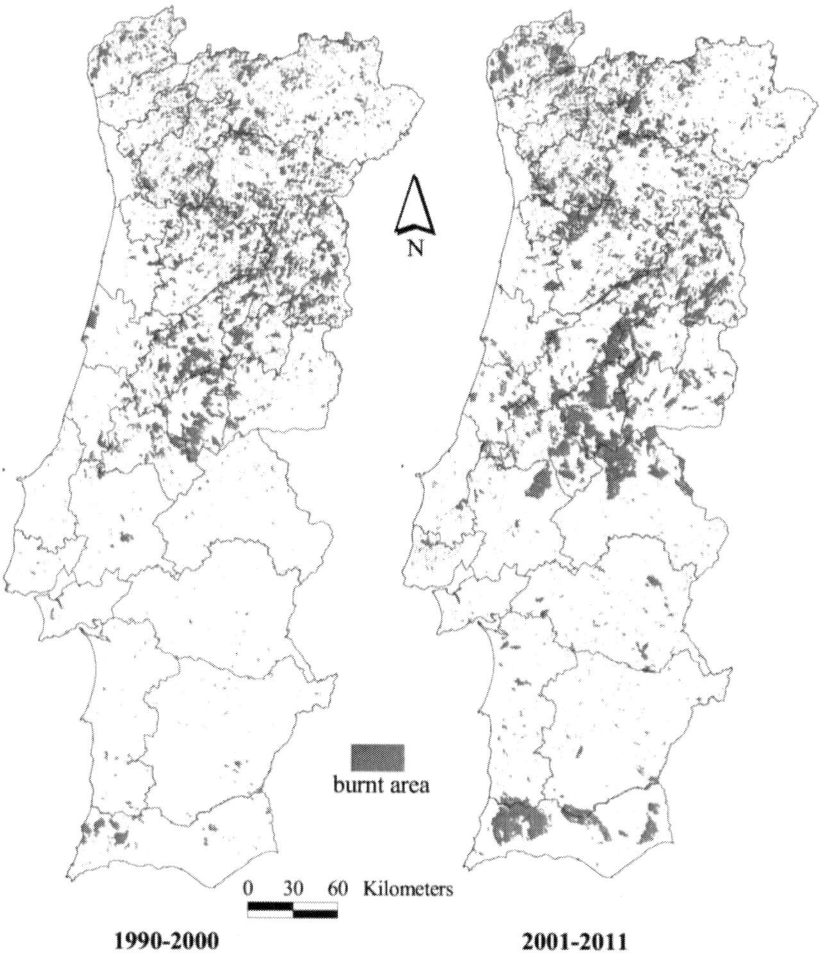

Source: Portuguese Government Forest Service.

Figure 8. Spatial incidence of forest fires with a burnt areas of more than 10 ha (1990-2011).

Recent socio-economic change (especially rural exodus) has resulted in the decline of traditional agriculture, facilitating the growth and consolidation of compact wood and shrub masses in abandoned agricultural land and, in turn, the spread of large fires. Other factors that explain the increasing number of forest fires in the Mediterranean context have also been identified, such as the transition to fossil fuels (which renders the use of biomass redundant in certain industrial and domestic contexts) and the new conservation and leisure functions of forests (which has openedthese areas up to urban populations avid for ''nature'' at the expense of traditional uses) (Badia et al., 2002).

Although Mediterranean vegetation is able to cope with fire (Trabaud, 1987; Pausas and Vallejo, 1999), alterations in the fire regime (in terms of greater fire recurrence or incidence of large events, for instance) can generate important consequences for Mediterranean ecosystems (Vazquez and Moreno, 2001; Lloret et al., 2003) and their plant diversity. Fire affects the biogeochemical properties of the soil, and the hydrology and geomorphology of the area, influencing the vegetation cover and, consequently, the characteristics of subsequent fires (Whelan, 1995).

The effects of fire on vegetation are very complex, not only because of the great complexity of Mediterranean ecosystems and the interactions with land-uses, but also because of the different responses to different type of fires and fire regimes (i.e., different intensities, seasonalities, recurrences and extent). At the landscape level, post-fire regeneration depends mainly on the initial vegetation (that is, plant traits of the initial species occurring on the site) and onsite environmental factors (climatic and terrain parameters) (Pausas and Vallejo, 1999). According to Silva and Catry (2006), the remarkable resistance of cork oak to fire gives it a considerable advantage compared to other tree species (with the possible exception of *Eucalyptus globulus*) , since much less effort is normally needed to rehabilitate the burned area. Proença (2009) also suggests that natural forest patches of broadleaved species (*Quercus robur* and *Ilex aquifolium*) are more resistant to fire than pine plantations. This is a great advantage which should be taken into consideration in reforestation policies.

Epidemic Diseases

The pinewood nematode, *Bursaphelenchus xylophilus*, is the causal agent of pine wilt disease, and is considered a quarantine pest by the European and Mediterranean Plant Protection Organization (EPPO, 2009). This nematode has been detected since 1999 in Portuguese forests south of Lisbon, affecting the maritime pine, *Pinus pinaster* Ait. (Mota et al., 1999). This species has great economic value, as it is the main coniferous species used in pine timber and resin production, both for export and domestic use. According to the last National Forest Inventory (2005/2006), *Pinus pinaster* occupies an area corresponding to 22.65% of the total forest area.

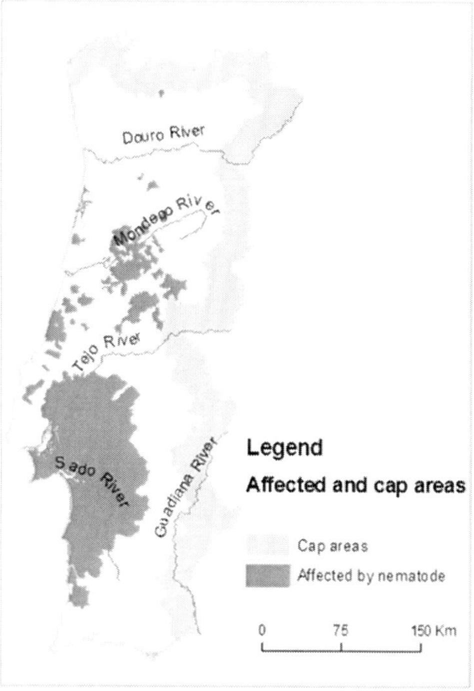

Figure 9. Up - areas affected by nematode and cap areas on mainland Portugal (Source: Portuguese Forest Service); below– maritime pine stand affected by nematode and invaded by *Acacia* sp.

Bursaphelenchus xylophilus originated in North America, spreading to Japan in the early 20th, and later into China, Korea and Taiwan. Trees of the genus *Pinus* are the main hosts for the pinewood nematode. The main mechanism for infection begins with the contact of the nematode with the trachea of its vector insects, mainly beetles of the genus *Monochamus* (Naves et al., 2007).

Physiological changes associated to pine wilt disease include decrease in photosynthesis, dysfunction of xylem and cortex parenchyma cells, production of phytotoxic substances and cambium destruction (Fukuda 1997; Rodrigues et al., 2010). Xylemdysfunction, causing disruption of water transport, is associated to aspiration of bordered pits (Ikeda and Kiyohara 1995), accumulation of volatile monoterpenes (Kuroda 1991), and cavitation and embolism in tracheids (Fukuda 1997; Ikeda 1996; Kuroda 1991).

The main symptoms of nematode infection consist of a general and rapid weakening of the tree with discoloration and total or partial drying of crowns (Figure 9). In the process, needles turn yellow and brown, starting with the younger ones, though they are nevertheless able to stay on the tree for a long period of time (external symptom). Symptom development is a process following at least a two-stage pattern (Fukuda 1997; Kuroda 1991; Shaheen et al. 1984).

Cavitation of the inner part of the xylem is an early-stage symptom. Atthe advanced stage, large-scale cavitation and embolism of the outermost xylem and cambial necrosis may cause the death of the tree (Fukuda 1997; Ikeda 1996). According to Kuroda (1991), the infected tree will die within 3 months of nematode invasion.

In order to circumvent the spread of the nematode pest, the Portuguese authorities have imposed regulatory measures to prevent its propagation (National Eradication Program for the Pinewood Nematode—Ministry of Agriculture, Sea, Environment and Regional Planning). However, due to insufficient implementation of the phytosanitary strategy, this nematode has spread rapidly to other regions of the country (Reva et al., 2011), mainly in the centre of Portugal.

A set of regulations have thus been enforced by the Portuguese authorities, establishing concepts such as “demarcated zones” (defined as areas where occurrence of nematode is acknowledged and which require effective intervention) and “cap zones” (defined as those encircling the former by 20 km) (Figure 9).

Alien Invasive Plant Species

With regard to exotic plants, over 550 species have been introduced into Portugal mainland. Approximately 8% of these have been classified as *invasive* species (Marchante et al., 2008; Almeida and Freitas, 2006). Among the organisms with higher negative impacts on the habitats, the *Acacia* Mill species are the most problematic.

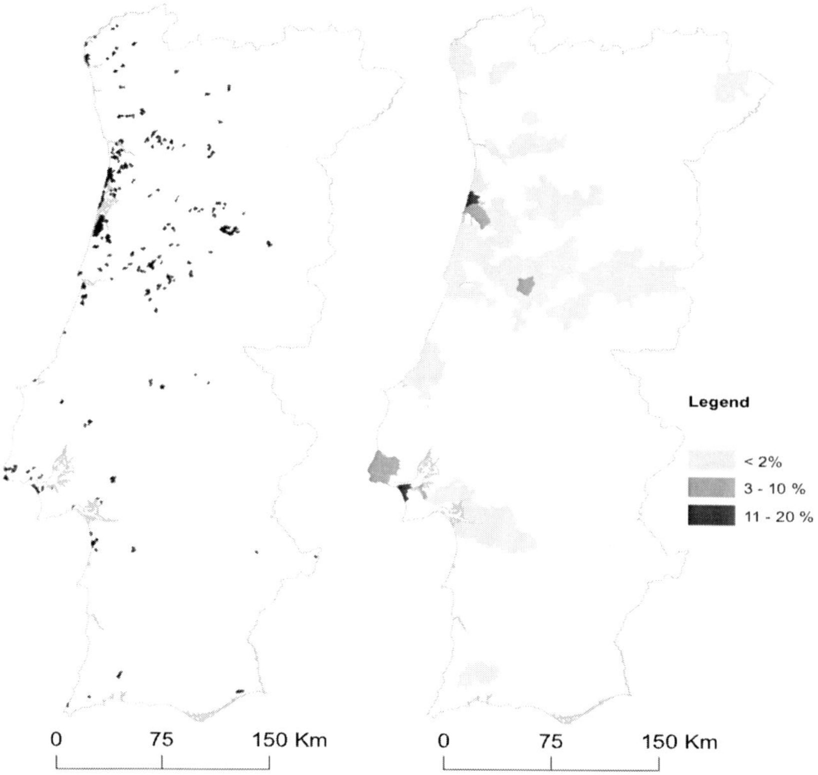

Figure 10. Presence of *Acacia*[2] species (*A.* melanoxylon, *A. dealbata*, *A. longifolia*): left – 1965 – dispersed occurrence (Source: Atlas Digital do Ambiente, CNA, 1977); right – 2005 - percentage of forest cover (DGRF, 2006).

[2] The published statistics are supported by different levels of taxonomic, organizational and spatial downscaling. In terms of taxonomy, data is aggregated at the genus level, so it is not possible to analyze responses at the species level.

Introduced for several reasons (ornamental, erosion control), some of these species have been considered high-risk (Marchante et al., 2008), a status determined by their current wide distribution, potential to invade new areas, and impacts on the habitats invaded (*Acacia cyclops* A. Cunn. Ex G. Don fil.; *Acacia longifolia* (Andrews) Willd.; *Acacia melanoxylon* R. BR., *Acacia retinodes* Schlecht, *Acacia saligna* (Labill.) H. L. Wendl.). Despite their lower occupancy area, the Australian cheesewood (*Pittosporum undulatum* Vent.) and the black locust (*Robinia pseudoacacia* L.) are two other tree alien invasive species that present a higher potential for invasion.

Acacia species have been introduced into different regions of the world for diverse purposes (see Kull et al., 2011). This was usually in order to bring about positive impacts, though the potential negative consequences for habitats were often neglected, such as the floristic impoverishment of communities (for a full revision of impacts see Le Maitre et al., 2011).

Their potential to invade new areas is associated to specific traits, which are common among invasive organisms. These include rapid growth rates and the capacity to set up a large and persistent seed bank, which give them an advantage in competitive interactions with native species after disturbance (Gibson et al., 2011).

In Portugal, Australian acacias are the most widely distributed alien species, and are present in different types of habitat. From sand dunes on the coast (*A. longifolia*) to siliceous mountain ranges (*A. dealbata*, *A. melanoxylon*), acacias registered an important increase in occupancy area after introduction.

Despite their specific attributes to dispersal and colonization, their use in forestry has contributed to their current wide distribution. In the 1960s, (Figure 10, left) the areas with acacias were mainly located in the western sectors of the territories north of the River Tejo, particularly mountain ranges and sand dune habitats in the centre of Portugal. In both territories, the species were introduced to control erosion, mainly because of the high disturbance exhibited by native communities.

The blue wattle (*A. dealbata*) is currently the species with the widest distribution and highest occupancy area. Because of its large ecological range (the limits of which are not set with precision in Portugal), a significant area of the territories north of the River Tejo have suitable conditions forits presence. As a considerable area is associated to the abandonment of traditional activities (agriculture, livestock, grazing) or disturbance by fire, such territories are very vulnerable invasion. At the municipality level there are some territories where stands dominated by acacia represent more than 10% of the area covered by forest (Figure 10).

Climate Change

Several authors have reported that of all bioclimatic regions, the Mediterranean area appears to be particularly vulnerable to global change (Sala et al., 2000, Lindner et al., 2010). As it functions as a transition zone between arid and humid regions of the world, it is especially sensitive to climate change (Scarascia-Mugnozza et al., 2000). Despite differences in magnitude, climate change scenarios predict an increase in temperature and significant changes in the precipitation patterns in the Mediterranean area. Precipitation may decrease by as much as 20%, particularly in summer, accompanied by an increase in the frequency and intensity of drought periods, affecting water resources, forestry and agriculture (Resco de Dios et al., 2007).

Higher temperatures and lower precipitation during summer will increase the evaporative demands of the atmosphere on ecosystems that can already transpire up to 80% of precipitation (Palahi et al. 2008). The IPCC 4th Assessment (2007) predicts an average temperature increase for the Mediterranean region of about 4-6°C by the end of the century, with 20-50% less rainfall during the summer months. Mediterranean conditions are expected to expand to new areas. Moreover, increased areas of unmanaged forest in the north, and deforestation and overexploitation in the south will increase the level of biotic (pests and disease) and abiotic (fires, droughts, etc.) risks, reducing the possibilities for adaptation to climate change.

The impacts of climate change on biodiversity are one of the strategic sectors identified in the recent proposal for a National Strategy on Adaptation to Climate Change. Studies indicate that throughout the 20th century the Portuguese climate has undergone three periods of change in mean temperature, with the most recent period (1976-2000) characterized by an accelerated increase in temperature. All the projected scenarios predict a significant increase in mean temperature and a reduction of precipitation, which will have a further impact on biological diversity.

Portugal is also one of the European Mediterranean countries where desertification brings the highest environmental risk (Sequeira, 1998a, b, c *in* MEA, 2004), particularly in connection with soil degradation caused by water erosion (Giordano, 1992 *in* MEA, 2004) and the degradation of surface and underground waters (GCID, 1998 *in* MEA, 2004). Portuguese ecosystems and agriculture are also particularly vulnerable to climate change (Santos et al., 2002 *in* MEA, 2004) due to the associated risks of fire and drought.

The predictions in terms of temperature and precipitation for the next century are in agreement with the trend detected in the last century (Santos and Miranda, 2006). The increase is clear in terms of temperature, particularly after 1970, but is less evident in terms of precipitation. Despite the general trend to decrease, there are different variations in terms of magnitude and direction between months. The decrease in winter and early spring (January, February, and March) is far more significant than the increase in October and December (see Santos and Miranda, 2006). Changes in precipitation patterns might be ecologically more significant than a simple reduction in totals. Moreover, projections also suggest an increase in extreme events, such as heat waves and heavy rain episodes, in terms of frequency and intensity (IPCC, 2007).

The vulnerability of forests to climate change depends essentially on three groups of factors: i) the attributes of the species (tolerance and potential to adapt); ii) the conditions of the habitat, and iii) the magnitude of the change.

The impacts can be assessed in terms of occupancy area, composition, structure and functional processes. The lack of data about such processes imposes limitations on our capacity to predict alterations associated to climate change, such as reproduction rates, dispersion and tolerance to pests and diseases. Moreover, it is difficult to identify the contribution of climate change among all the possible factors.

Changes in phenological processes, more easily detected on highly disintegrated time scales, have been reported for some species in the Mediterranean region (Peñuelas et al., 2002). But other changes are difficult to detect. Because of the longevity of trees (decades to centuries), the assessment of the impacts from climate change on forests is often associated to an analysis of alterations in disturbance regimes (such as the frequency and intensity of fire, drought, alien species invasion, plagues of insects and pathogens, windstorms, ice storms). In fact, the action of such disturbances might overshadow the direct influence of climate change (Dale et al. 2001). Usually the temporal scale used to predict changes cannot show the natural migration of trees as result of climate change, as this occurs on much larger time scales.

Because of decreasing equilibrium in terms of ecological conditions, populations will be more vulnerable to pests and diseases at the distribution edges. But impacts could also occur in areas where species are in equilibrium with ecological conditions, and these situations could be associated to higher vulnerability to disturbance by fire, enhanced by the higher probability of extreme climatic conditions, such as summer heat waves, which are responsible for large areas affected by wildfire in Portugal (e.g. summer 2003). Moreover, the contribution of disturbance by fire may become particularly significant because of the dominance of fire-prone communities in the landscape (Almeida et al., 2009).

The impacts of climate change might be reinforced by synergetic contributions from other factors, such as the loss and fragmentation of habitat. It is expected that the high degree of disturbance exhibited by the native forests could exacerbate the impact of climate change. The projected reduction in precipitation and rise in temperature will bring consequences in terms of a wood's regeneration rate, particularly in the driest parts of the territory (Douro valley, inland south), where the woods, despite the tolerance to dry conditions, present a high level of disturbance.

In terms of forest productivity, a general decline is predicted, mostly due to an increase in the extent and magnitude of the climatic water deficit (Pereira et al., 2002), resulting from the decreasing precipitation, higher frequency of heavy rain episodes and rising temperature (Santos and Miranda, 2006). Such changes are likely to bring more profound changes at the edges of the potential deciduous oak forest area, compared to the evergreen. Current levels might be maintained or even increase in the northwest sector as a result of rising temperatures (Pereira et al., 2006). This might extend the growth period and expand the suitable area where the pedunculate forests are dominant, due to a rise in the minimum temperature. But this scenario of higher productivity and expansion must be calibrated by a context of lower precipitation, especially during spring, which might constitute a serious constraint determining a loss of suitable area.

Despite the lower vulnerability of evergreen oaks to dryness, the current soil degradation and the forests decrease in the south will probably reinforce the negative impacts associated to a larger dry season. Such a scenario will probably have higher impacts on the cork oak forests, not only in terms of distribution area but also with regard to their vulnerability to pests and diseases (Pereira et al., 2002). The current forests might persist, considering the resilience of evergreen oaks. But the regeneration of such woods might be compromised because of the higher vulnerability of juveniles to summer dryness. In fact, increased warmth and dryness may be already contributing to the decline of the evergreen forests. The higher levels of tree mortality are interpreted as a consequence of greater vulnerability to pathogens and diseases (Resco de Dios et al., 2007).

As water-demanding forests, the eucalyptus and maritime pine forests will register a significant reduction in the area where cultivation is profitable (Pereira et al., 2006), despite the potential increase in suitable areas at higher altitudes in the mountain ranges north of the Douro valley. It is predicted that climate change might restrict the area of potential exploitation to the western sectors of the territory north of the Tejo basin. (Pereira et al, 2006). The higher frequency of fires might present a huge obstacle to the exploitation of such forests.

In terms of distribution, it is also predicted that the current forests will shift northward and westward (Pereira et al., 2002), and that there will be a reduction in the distribution area of higher water-demanding forests (e.g. pedunculate oak forest) and their replacement by evergreen oak forests.

CONCLUSION

There is a significant difference between the composition and occupancy area of the potential forests and the current woodlands that dominate the landscape in mainland Portugal. Such changes are much more significant in the territories north of the River Tejo. In fact, native oak forests, particularly deciduous and semi-deciduous varieties, are currently restricted to small and highly disturbed patches as a result ofcenturies of cutting, grazing, wildfires and afforestation, reinforced by the use of exotic species, activities that leave their fingerprint on the composition and structure of the remnant stands.

While the changes in the northern territories are of a floristic nature, in the south, the changes are mainly in the structure of the native forests, which are dominated by evergreen oaks (cork oak and holm oak). Cork oak stands are associated with a remarkable biodiversity and constitute unique ecosystems, which are recognized for their ecological value. This ecological importance was taken into account in the classification of cork oak stands as protected habitats in the framework of the Natura 2000 Network, established by the European Union (Directive no. 92/43/CEE) since 1993. This is the case of Habitat 6310 – *Montado*s of evergreen *Quercus* spp. and Habitat 9330 – *Quercus suber* forests (Silva and Catry, 2006).

Woodlands have been increasing, mostlyl due to intensive forestation programmes carried out during the last century, based on pine (*Pinus pinaster)* and eucalyptus (*Eucalyptus globulus).*

Thus, the Portuguese forest is presently dominated by eucalyptus and pine plantations, which have affected the distribution and condition of the native oak forests, either because plantations replace oak forests or occupy land where deciduous oak forest could regenerate or be planted. Fire disturbance also compromises the possibility of future forest regeneration or causes the degradation of existing forests.

In fact, forest fires are, nowadays, the major threat to the sustainability of Portuguese forests, an issue that becomes even more relevant within the context of climate change. These environmental changes might reinforce the vulnerability of plants to diseases, because climate change will affect the pathogen, host and the interaction between them, resulting in changes in the disease's impact. Predicted climate change might also contribute to increase the suitable area and even enhance the vulnerability of territories to invasion by alien invasive species.

REFERENCES

AFN (Autoridade Florestal Nacional). (2007). Inquérito Florestal Nacional e IFN 1995e2005. Lisboa: Autoridade Florestal Nacional (AFN).

Almeida, A.C., Nunes, A., Figueiredo, A. (2009). Mudanças no uso e cobertura vegetal no Interior Centro e Norte de Portugal. Imprensa da Universidade de Coimbra, 99 p.

Almeida, J.D. and Freitas, H. (2006). Exotic naturalised flora of continental Portugal – a reassessment. *Botanica Complutensis* 30: 117-130.

Almeida, J.D. (1999). Flora exótica subespontânea de Portugal continental (plantas vasculares). 2.ª edição. Catálogo das plantas vasculares exóticas que ocorrem subespontâneas em Portugal continental e compilação de informações sobre estas plantas. Dissertação de Mestrado. Faculdade de Ciências e Tecnologia da Universidade de Coimbra.

APA (2008). Relatório do Estado do Ambiente 2007. Agência Portuguesa do Ambiente. Amadora.

Badia, A., Saurí, D., Cerdan, R., Llurdés, J.C. (2002). Causality and management of forest fires in Mediterranean environments: an example from Catalonia. *Environmental Hazards*, 4, 23- 32.

Baptista, F. O. (1993). In A política agrária do Estado Novo. Porto: Afrontamento.

Bingre, P. and Damasceno, P. (2007). Biologia e ecologia das florestas de carvalho-português. In J. S. Silva (Ed.), *Os Carvalhais. Um Património a Conservar* (Vol. 2, pp. 15-45). Col. Árvores e Florestas de Portugal,

Lisboa: Fundação Luso-Americana, Publico and Liga para a Proteção da Natureza.

Caldas, E. C. (1978). *A agricultura portuguesa no limiar da Reforma Agrária.* Fund. Calouste Gulbenkian, Centro de Estudos de Economia Agrária, Oeiras, 259 p.

Capelo, J. and Catry, F. (2007a). A distribuição do carvalho-português. In J. S. Silva (Ed.), Os Carvalhais. Um Património a Conservar (Vol. 2, pp. 83-94). Col. Árvores e Florestas de Portugal, Lisboa: Fundação Luso-Americana, Publico and Liga para a Proteção da Natureza.

Capelo, J. and Catry, F. (2007b). A distribuição do sobreiro em Portugal. In J. S. Silva (Ed.), *Os Montados. Muito para além das árvores.* (Vol. 3, pp. 107 - 113). Col. Árvores e Florestas de Portugal, Lisboa: Fundação Luso-Americana, Publico and Liga para a Proteção da Natureza.

Carvalho, J., Alves, P.C., Grosso-Silva, J. M., and Santos, T. M.d. (2007a). Biologia e ecologia das florestas de carvalho-negral. In J. S. Silva (Ed.), *Os Carvalhais. Um Património a Conservar* (Vol. 2, pp. 99-120). Col. Árvores e Florestas de Portugal, Lisboa: Fundação Luso-Americana, Publico and Liga para a Proteção da Natureza.

Carvalho, J., Alves, P. C., Grosso-Silva, J. M., Santos, T. M. d. (2007b). Biologia e ecologia do carvalho-roble. In J. S. Silva (Ed.), *Os Carvalhais. Um Património a Conservar* (Vol. 2, pp. 185 - 201). Col. Árvores e Florestas de Portugal, Lisboa: Fundação Luso-Americana, Publico and Liga para a Proteção da Natureza.

Carvalho, J. and Devy-Vareta, N. (2007). O carvalho-negral ao longo da história. In J. S. Silva (Ed.), Os Carvalhais. Um Património a Conservar (Vol. 2, pp. 137-151). Col. Árvores e Florestas de Portugal, Lisboa: Fundação Luso-Americana, Publico and Liga para a Proteção da Natureza.

Castro, E.B., González, M.A.C., Tenorio, M. C., Bombin, R. E., Antón, M. G., Fuster, M. G., Ollero, H. S. (2005). Los Bosques Ibéricos. Una Interpretación Geobotánica (4 ed.). Barcelona: Editorial Planeta, 597 pp.

Castroviejo, S., Laínz, M., González, G. L., Montserrat, P., Garmendia, F. M., Paiva, J., and Villar, L. (Eds.) (1986-2012). *Flora Iberica. Plantas vasculares de la Península Iberica e Islas Baleares.* Madrid: Real Jardín Botánico, C.S.I.C.

Coelho, I. S. (2003). Propriedade da Terra e Política Florestal em Portugal. *Silva Lusitana* 11(2), 185-199.

Coelho, I. (1996). O Montado, a economia e o desenvolvimento do Alentejo. *Silva Lusitana* 4: 39–48.

Costa, A., and Pereira, H. (2007). Montados e Sobreirais: uma espécies, duas perspectivas. In J. S. Silva (Ed.), *Os Montados. Muito para além das árvores.* (Vol. 3, pp. 17-37). Col. Árvores e Florestas de Portugal, Lisboa: Fundação Luso-Americana, Publico and Liga para a Proteção da Natureza.

Costa, J. C., Aguiar, C., Capelo, J., Lousa, M., Neto, C. (1998). Biogeografia de Portugal Continental. *Querceta* 0, 5-56.

Croitoru, L. (2007). How much are Mediterranean forests worth? *Forest Policy and Economics* 9: 536 – 545.

Dale, V.H., Joyce, L.A., McNulty, S., Neilson, R.P., Ayres, M.P., Flannigan, M.D., Wotton, B. (2001). Climate Change and Forest Disturbances. *BioScience* 51(9), 723-734.

Devy-Vareta, D. (1985). Problèmes de la forêt au Portugal. *Revue Géographique des Pyrénées et du Sud-ouest*, Tomo 51, Fasc.3.

Devy-Vareta, D. (1986). Para uma Geografia histórica da floresta portuguesa; do declínio das matas medievais à política florestal do Renascimento (séc. XV e XVI). *Revista da Faculdade de Letras, Geografia,* 1ª Série, Vol. I, Porto, pp. 5-39.

DGDR/IDRHa. (1998). Medidas agro-ambientais e Novo programa de aplicação em Portugal Continental. Épocas de 1990 e 1999. Lisboa: DGDR.

DGDR/IDRHa. (2004). Medidas agro-ambientais. Lisboa: RURIS Plano de Desenvolvimento Rural, DGDR.

DGRF (Direcção Geral dos Recursos Florestais). (2006). Incêndios Florestais e Relatório 2005. Lisboa, Portugal: Divisão de Defesa da Floresta contra Incêndios, Direcção Geral dos Recursos Florestais.

DGRF/AFN. (2006). Estratégia Nacional das Florestas, ex-Direcção Geral dos Recursos Florestais (DGRF). Lisboa.

Doerr S.H., Shakesby, R.A., Walsh, R.P.D. (2000). Soil water repellency: its causes, characteristics and hydro-geomorphological significance. *Earth-Science Reviews* 51:33-65.

EPPO (2009). EPPO A1 and A2 Lists of pests recommended for regulation as quarantine pests. http://www.eppo.org/QUARANTINE/quarantine.htm. Accessed April 2010.

EPPO (2009). EPPO A1 and A2 Lists of pests recommended for regulation as quarantine pests. http://www.eppo.org/QUARANTINE/quarantine.htm. Accessed April 2010

Fabião, A., Oliveira, Â., Correia, A.V. (2007). Conservação, regeneração e exploração dos carvalhais. In J. S. Silva (Ed.), *Os Carvalhais. Um Património a Conservar* (Vol. 2, pp. 67 - 81). Col. Árvores e Florestas de

Portugal, Lisboa: Fundação Luso-Americana, Publico and Liga para a Proteção da Natureza.

Figueiredo, A. (2005). *Vegetação natural potencial num contexto de elevada perturbação: o caso da depressão de Chaves.* Master MSc, Universidade de Coimbra, Coimbra, 184 pp.

Franco, J. d. A. (1990). Quercus. In S. Castroviejo, C. Aedo, M. Laínz, F. M. Garmendia, G. N. Feliner, J. paiva and C. Benedí (Eds.), *Flora Iberica* (Vol. II - *Fagaceae*, pp. 15-36). Madrid: Real Jardín Botánico, CSIC.

Franco, J. d. A., and Afonso, M. L. R. (1998). *Nova Flora de Portugal (Continente e Açores)* (Vol. III (2) - Graminae). Lisboa: Escolar Editora.

Fukuda, K. (1997). Physiological process of the symptom development and resistance mechanism in pine wilt disease. *J. For. Res.* 2:171– 181.

Géhu, J. M., and Rivas-Martínez, S. (1981). Notions fondamentales de phytosociologie. In H. Dierschke (Ed.), *Syntaxonomie. Berichte der Internationalen Symposien der Internationalen Vereinigung Vegetationskunde.* (pp. 5-33). Vaduz.

Gibson, M.R., Richardson, D.M., Marchante, E., Marchante, H., Rodger, J. G., Stone, G. N., Wilson, J.R.U. (2011). Reproductive biology of Australian acacias: important mediator of invasiveness? *Diversity and Distributions* 17(5): 911-933.

Godinho-Ferreira, P., Azevedo, A., Rego, F. (2005). Carta da Tipologia Florestal de Portugal Continental. *Silva Lusitana* 13(1): 1-34.

Gonzalez, J. R. and Pukkala, T. (2007). Characterization of forest fires in Catalonia (North-east Spain). *European Journal of Forest Research* 126 (3): 421-429.

Hobbs R.J, Richardson, D.M, Davis G.W. (1995). Mediterranean-type ecosystems: opportunities and constraints for studying the function of biodiversity. In: Richardson D.M, Davis G.W. (eds.) *Mediterranean-type ecosystems. The function of biodiversity, Ecological Studies,* Vol. 109, Springer-Verlag, New York, pp. 1-42.

Ikeda, T. (1996). Xylem dysfunction in Bursaphelenchus xylophilusinfected Pinus thunbergii in relation to xylem cavitation and water status. *Ann. Phytopatol. Soc. Jpn* 62: 554–558.

Ikeda, T., Kiyohara, T. (1995). Water relations, xylem embolism and histological features of Pinus thunbergii inoculated with virulent or avirulent pine wood nematode Bursaphelenchus xilophilus. *J. Exp. Bot.* 46 (285): 441–449.

IPCC. (2007). Climate Change 2007: The Physical Science Basis. Contribution of Working Group I to the Fourth Assessment Report of the

Intergovernmental Panel on Climate Change. In S. Solomon, D. Qin, M. Manning, M. Marquis, K. Averyt, M. M. B. Tignor, H. L. Miller and Z. Chen (Eds.), (pp. 996). Cambridge, United Kingdom and New York, NY, USA: IPCC.

Kull, C. A., Shackleton, C. M., Cunningham, P. J., Ducatillon, C., Dufour-Dror, J.-M., Esler, K. J., Zylstra, M. J. (2011). Adoption, use and perception of Australian acacias around the world. *Diversity and Distributions* 17(5): 822-836.

Kuroda, K. (1991). Mechanism of cavitation development in the pine wilt disease. *Eur. J. For. Path* 21:82–89.

Leite, F. F., Gonçalves, A. B., and Vieira, A. (2011). The recurrence interval of forest fires in Cabeço da Vaca (Cabreira Mountaindnorthwest of Portugal). *Environmental Research* 111 : 215-221.

Le Maitre, D. C., Gaertner, M., Marchante, E., Ens, E.-J., Holmes, P. M., Pauchard, A., Richardson, D. M. (2011). Impacts of invasive Australian acacias: implications for management and restoration. *Diversity and Distributions* 17(5): 1015-1029.

Lloret, F., Pausas, J.G., Vila, M. (2003). Responses of Mediterranean plant species to different fire frequencies in Garraf Natural Park (Catalonia, Spain): Field observations and modeling predictions. *Plant Ecology* 167: 223-235.

Marchante, H., Marchante, E., Freitas, H. (2005). Invasive plant species in Portugal: an overview. In: *International Workshop on Invasive Plants in Mediterranean Type Regions of the World* (ed. S. Brunel), Montpellier, France. Council of Europe Publishing, pp. 99-103.

Marchante, H, Campelo, F., Freitas, H. (2006). Avaliação do impacto de *Acacia longifolia* [Andrews] Willd. sobre a diversidade vegetal dos sistemas dunares (Dunas de S. Jacinto e Palheirão). In: Rodrigues, L.; Reino, L.; Gordinho, L.O and Freitas, H. (eds.) Actas do 1º Simpósio sobre Espécies Exóticas: Introduções, Causas e Consequências. Pp. 67-71. Liga para a Protecção da Natureza, Lisboa.

Marchante, E., Freitas, H., Marchante, H. (Eds.). (2008). *Guia Prático para a Identificação de Plantas Invasoras de Portugal Continental.* Coimbra: Imprensa da Universidade de Coimbra.

MEA (2004). Pereira, H.M, T. Domingos, and L. Vicente (editors). Portugal Millennium Ecosystem Assessment: State of the Assessment Report. Centro de Biologia Ambiental, Faculdade de Ciências da Universidade de Lisboa.

Mendes, A. (2002) A economia do sector da cortiça em Portugal. Evolução das actividades de produção e transformação ao longo dos séculos XIX e XX, in: *XXII Encontro da Associação Portuguesa de História Económica e Social* (Aveiro: Universidade de Aveiro), 268 pp.

Mendes, A. M. S. C. (2007). O sobreiro ao longo dos tempos. In J. S. Silva (Ed.), *Os Montados. Muito para além das árvores.* (Vol. 3, pp. 77 - 106). Col. Árvores e Florestas de Portugal, Lisboa: Fundação Luso-Americana, Publico and Liga para a Proteção da Natureza.

Merlo M. and Rojas Briales E. (2000). Public goods and externalities linked to Mediterranean forests: economic nature and policy. *Land Use Policy* 17: 197-208.

Monke, E., Avillez, F., Pearson, S., Gaetano, M. (1998). Evaluation of small farm agriculture, chapter 2. In E. Monke, F. Avillez, and S. Pearson (Eds.), Small farm agriculture in Southern Europe e CAP reform and structural change. Aldershot: Ashgate Editions.

Moreira, F., Viedma, O., Arianoutsou, M., Curt, T., Koutsias, N., Rigolot, E., et al. (2011). Landscape e wildfire interactions in southern Europe: implications for landscape management. *Journal of Environmental Management* 92: 2389-2402.

Mota, M. M., Braasch, H., Bravo, M.A., Penas, A.C., Burgermeister, W., Metge, K., Sousa, E. (1999). First report of Bursaphelenchus xylophilus in Portugal and Europe. *Nematology* 1:727–734.

Mouillot F. Rambal S., Joffre R. (2002). Simulating climate change impacts on fire frequency and vegetation dynamics in a Mediterranean-type ecosystem. *Global Change Biology* 8 (5): 423–437.

Natividade, J. (1950). *A Subericultura* (Lisbon: Direcção-Geral dos Recursos Florestais e Aquícolas).

Naveh, Z. (1987). Landscape ecology, management and conservation of European and Levant Mediterranean uplands. In: Tenhunen J.D, Catarino F.M, Lange O.L, Oechel W.C. (eds) *Plant Response to Stress. Functional Analysis in Mediterranean Ecosystems. NATO ASI series* Vol. G.15, Springer-Verlag, Berlin-Heidelberg-New York. pp. 641-657.

Naves, P., Camacho, S., Sousa, E., Quartau, J. (2007). Transmission of the pine wood nematode Bursaphelenchus xylophilus through oviposition activity of Monochamus galloprovincialis (Coleoptera: Cerambycidae). *Entomol Fenn* 18:193–198.

Noss, R.F., Franklin, J. F., Baker, W. L., Schoennagel, T., and Moyle, P. B. (2006). Managing fire-prone forests in the western United States. *Frontiers in Ecology and the Environment* 4: 481-487.

Nunes, A. (2007). Abandono do espaço agrícola na "Beira Transmontana": Extensão, causas e efeitos ambientais. Diss. Doutoramento, FLUC, Coimbra, 317 p.

Nunes, A. N. (2012). Regional variability and driving forces behind forest fires in Portugal: an overview of the last three decades (1980-2009). *Applied Geography* 34: 576-586.

Nunes, A. N., Coelho, C. O. A., Almeida, A. C., Figueiredo, A. (2010). Soil erosion and hydrological response to land abandonment in a central Inland area of Portugal. *Land Degradation and Development*, 21(3): 260-273.

Nunes, C., Vasconcelos, M., Pereira, J., Dasgupta, N., Alldredge, R. and Rego, F. (2005). Land cover type and fire in Portugal: do fires burn land cover selectively? *Landscape Ecology* 20: 661–673.

Paiva, J. (2007a). O carvalho-português na história e na cultura. In J. S. Silva (Ed.), Os Carvalhais. Um Património a Conservar (Vol. 2, pp. 55-66). Col. Árvores e Florestas de Portugal, Lisboa: Fundação Luso-Americana, Publico and Liga para a Proteção da Natureza.

Paiva, J. (2007b). O carvalho-roble na história e na cultura. In J. S. Silva (Ed.), Os Carvalhais. Um Património a Conservar (Vol. 2, pp. 211-228). Col. Árvores e Florestas de Portugal, Lisboa: Fundação Luso-Americana, Publico and Liga para a Proteção da Natureza.

Pausas J.G. and Fernández-Muñoz S. 2012. Fire regime changes in the Western Mediterranean Basin: from fuel-limited to drought-driven fire regime. *Climatic Change* 110: 215-226.

Pausas, J., and Vallejo, R. (1999). The role of fire in European Mediterranean ecosystems. In E. Chuvieco (Ed.), Remote sensing of large wildfires in the European Mediterranean basin (pp. 3-16). Berlin: Springer-Verlag.

Peñuelas, J., Filella, I., and Comas, P. (2002). Changed plant and animal life cycles from 1952 to 2000 in the Mediterranean region. *Global Change Biology* 8(6): 531-544.

Pereira, J. S., Correia, A. V., Correia, A. P., Branco, M., Bugalho, M., Caldeira, M. C., Vasconcelos, M. J. (2002). Forests and Biodivesity. In F. D. Santos, K. Forbes and R. Moita (Eds.), *Climate Change in Portugal. Scenarios, Impacts and Adaptation Measures - SIAM Project* (pp. 454). Lisboa: Gradiva.

Pereira, P. and Fonseca, M. (2003) Nature vs. nurture: the making of the *montado* ecosystem. *Conservation Ecology* 7(3): 7.

Pereira, J. S., Correia, A. V., Correia, A. P., Ferreira, M. T., Onofre, N., Freitas, H., and Godinho, F. (2006). Forestas e Biodiversidade. In F. D. Santos and P. Miranda (Eds.), *Alterações Climáticas em Portugal.*

Cenários, Impactos e Medidas de Adaptação - Projecto SIAM II (pp. 506). Lisboa: Gradiva.

Proença, V.A.M. (2009). *Galicio-Portuguese oak forest of Quercus robur and Quercus pyrenaica: biodiversity patterns and forest response to fire*. PhD, Faculdade de Ciências da Universidade de Lisboa, Portugal.

Radich, M. C. AND Alves, A. A. M. (2000). *Dois séculos da floresta em Portugal*. Ed. CELPA, Lisboa, 226 pp.

Rego, F. C. (2001). *Florestas Públicas*. Ministério da Agricultura, do Desenvolvimento Rural e das Pescas, Lisboa, 105 p.

Reva V·, Fonseca L. , Lousada J. L. , Abrantes I., Viegas D. X. (2011). Impact of the pinewood nematode, Bursaphelenchus xylophilus, on gross calorific value and chemical composition of Pinus pinaster woody biomass. *European Journal of Forest Research*, Springer-Verlag.

Rivas-Martinez, S. (2004). Global Bioclimatics (Clasificación bioclimática de la Tierra). Retrieved from http://www.globalbioclimatics.org/book/bioc/global_bioclimatics_2.htm - 2d

Rodrigues A., Casquilho M., Oliveira H, Bordado J (2010). A statistical analysis of the impact of nematode attack symptomatology on the mechanical behaviour of Pinus pinaster Ait. wood. *Eur J Forest Res* 129(2):145–153.

Santos, F. D., and Miranda, P. (Eds.) (2006). *Alterações climáticas em Portugal. Cenários, impactos e medidas de adaptação* (1st ed.). Lisboa: Gradiva.

Shaheen, F., Winter, R.E.K, Bolla, R.I. (1984). Phytotoxin production in the Bursaphelenchus xylophilus-infected Pinus Sylvestris. *Jpn J Nematol* 16(1):57–6.

Silva, J. S. (Ed.). (2007). Os Carvalhais. Um Património a Conservar (Vol. 2). Lisboa: Público, Fundação Luso-Americana para o Desenvolvimento and Liga para a Proteção da Natureza.

Silva, J. S. and Catry, F. (2006) - Forest fires in cork oak (*Quercus suber* L.) stands in Portugal. *International Journal of Environmental Studies* 63 (3): 235–257.

Soares, J., L., Leal, P. Canaveira, F. Goes Fialho, A. (2007). Porquê cultivar o eucalipto? Pages 185–219 *in* Público comunicação Social, SA and Fundação Luso-Americana para o Desenvolvimento editors. *Árvores e Florestas de Portugal.* Pinhais e Eucaliptais – A floresta cultivada. Lisbon, Portugal.

Trabaud, L. (1987): Dynamics after fire of sclerophyllous plant communities in the mediterranean basin. *Ecología Mediterránea,* 13 (4): 25-37.

Vazquez A, and Moreno JM (2001). Spatial distribution of forest fires in Sierra de Gredos (Central Spain). *Forest Ecology and Management* 147: 55-65.

Whelan, R.J. (1995). The ecology of fire. Cambridge University Press, Cambridge, UK

In: Woodlands
Editor: Lazaro Manzanares
ISBN: 978-1-62618-177-9

Chapter 2

THE ROLE AND DIVERSITY OF SOIL FAUNA IN DIFFERENT WOODLANDS

Cristina Menta[1,*], Luis G. García-Montero[2], Federica D. Conti[1] and Stefania Pinto[1]

[1]Department of Bioscience, University of Parma, Parma, Italy
[2]Department of Forestry Engineering, E.T.S.I. Montes, Universidad Politécnica de Madrid (UPM, Technical University of Madrid), Ciudad Universitaria, Madrid, Spain

ABSTRACT

The increasing anthropic pressure on the environment is leading, in most parts of the world, to a rapid change in land use and an intensification of agricultural activities, in many cases at the expense of forest ecosystems.

These processes often result in soil degradation and a consequent loss of soil quality and soil biodiversity. Soil is one of the most important reservoirs of biodiversity. Within it is found the habitat of the roots of plants, microorganisms such as viruses, bacteria, fungi and algae and numerous invertebrates and vertebrates.

It reflects ecosystem metabolism since all the bio-geo-chemical processses of the different ecosystem components are combined within it. Organic matter degradation and nutrient cycling in the belowground ecosystems are regulated by very complex biotic and abiotic interactions,

[*] Corresponding author: cristina.menta@unipr.it.

where soil microflora and soil fauna directly and indirectly affect soil functioning. Soil fauna play an essential role in several soil ecosystem functions; furthermore, it is often used to provide soil quality indicators. In woodland ecosystems, plant composition, bearing the characteristics of the litter and the rhizosphere, affects both edaphic community structure and the dynamics of decomposition and recycling. On the other hand, edaphic community structure affects the decomposition of organic matter and the formation of the humus profile. In fact, it has been shown that soil invertebrates improve the soil fertility and productivity of forest ecosystems.

INTRODUCTION

The conservation of habitats and biodiversity must become one of the major goals of sustainable politics. Although global biodiversity was one of the focal points of the Rio Conference, in the 1990s virtually no attention was paid to activities for the conservation of soil communities.

However, with the New Millennium, the conservation of soil biodiversity has become an important aim of international environmental policies, as highlighted in the EU Soil Thematic Strategy (2006), the Biodiversity Action Plan for Agriculture (EU 2001), the Kiev Resolution on Biodiversity (EU/ECE 2003) and afterwards in the Message from Malahide (EU 2004), which lay down the goals of the 2010 Countdown. Soil is one of the most fundamental components for supporting life on Earth. Jeffery et al. (2010) and Menta (2012) highlight that the processes occurring within soil (most of which are driven by its living organisms) perform ecosystem and global functions that help maintain life above ground.

Moreover, soil delivers numerous services, ranging from providing the food we eat to filtering and cleaning the water we drink. It is used as a platform for building, it provides vital products such as antibiotics, and it contains an archive of our cultural heritage in the form of archaeological sites. However, life within the soil is hidden and often suffers from being 'out of sight and out of mind' (Blum, 1993; Ebel, 2007).

In effect often not enough importance is given to the fact that soil quality and its protection contribute significantly to preserving the quality of life, and that the nutrition and health of humans and animals cannot be separated from the quality of the soil.

SOIL FAUNA IN WOOD ECOSYSTEMS

The organisms that populate the soil almost all belong to the taxonomic groupings of prokaryotes and eukaryotes. Some of these taxa include forms present almost exclusively in the soil while others include forms present in both the soil and in the aboveground environment.

Soil biota are thought to harbour a large part of the world's biodiversity and to govern processes that are regarded as globally important components in the recycling of organic matter, energy and nutrients. Moreover, they are also key players in several supporting and regulating ecosystem services (Jeffery et al., 2010). Furthermore, they are key components of soil food webs. Their influence is so high that Ghilarov (1978) was able to state the general principle that the more similar the edaphic fauna of two soils are, the more these soils will be genetically similar in terms of their hydrobiology, chemistry and ecological conditions (Giordano, 1999). Rough estimates of soil biodiversity indicate several thousand invertebrate species per site, as well as the relatively unknown levels of microbial and protozoan diversity. The organic substance living in the soil includes microflora (bacteria, actinomycetes, fungi and algae), soil fauna, represented by organisms such as nematodes, springtails and earthworms, and finally, the roots of plants. The number of species, and the composition and the biological diversity of a soil depend on numerous factors such as the chemico-physical characteristics of the soil, the humidity, the temperature, the depth, the nutrient content and availability, the vegetational component, anthropic activity, etc. Generally, biodiversity tends to be greater in forest areas and permanent meadows, followed by cultivated areas, which are generally poorer in terms of species. The importance of soil organisms, especially microflora, lies in their capacity to give rise to new sources of energy transforming the residues of dead plants and animals into inorganic compounds or into neosynthesising organic compounds which are thus returned into the lifecycle of the ecosystem.

Soil ecosystems generally contain a large variety of animals, such as protozoans, nematodes, microarthropods such as mites and collembola, small insects, myriapods and enchytraeids and earthworms. In addition, a large number of meso- and macrofauna species (mainly arthropods such as beetles, spiders, diplopods, chilopods and pseudoscorpions, as well as snails) live in the uppermost soil layers, the soil surface and the litter layer.

Despite several decades of soil biological studies it is still very difficult to provide average abundance and biomass values for soil invertebrates. This is partly caused by their high variability in both time and space, as well as by differences in the sampling methods used (Jeffery et al., 2010). It is also well established that the biomass and abundance of different soil fauna groups fluctuate strongly in time and space. These spatio-temporal patterns may be a consequence of changes in microclimatic soil properties or due to the phenological characteristics of the organisms themselves (Taylor et al., 2010). As a result community composition undergoes constant shifts in the abundance or biomass of individual fauna groups. These fluctuations in community composition are likely to affect ecosystem processes not only directly but also indirectly, as changes in one soil fauna group will inevitably induce changes in other groups via a range of either trophic or non-trophic interactions (Taylor et al., 2010).

Soil fauna are highly variable and the majority are also highly adaptable with regard to their feeding strategies, ranging from herbivores to omnivores and even including carnivores. Depending on the available food sources many soil fauna are able to change their feeding strategies to a greater or lesser extent with many carnivorous species able to feed on dead organic matter in times of low food availability. The dominance of functional groups within a community can strongly depend on interactions like predation and competition but also on facilitation that affects the population dynamics or resource utilization of single species or groups (Taylor et al., 2010). In fact, the interactions between soil fauna are numerous, complex and varied. As well as the predator / prey relationships and in some instances parasitism, commensalism also occurs.

The degree of interaction between soil organisms and the soil itself can be highly variable among taxa and dependent on the part of the life cycle that is spent in the soil (Wallwork, 1970). In particular, in relation to this aspect, combined with the morphological adaptations and the ecological functions of organisms, it is possible to classify soil fauna into four main groups: temporarily inactive geophiles, temporarily active geophiles, periodical geophiles, and geobionts. These groupings do not have any taxonomical significance but are useful when studying the life strategies of soil invertebrates. Temporarily inactive geophiles are organisms that live in the soil for only certain phases of their life, such as to overwinter or to undergo metamorphosis, when protection from climatic instability is more necessary. Due to their relative inactivity, the organisms belonging to this group have a weak influence on the ecological function of soil, although they can be important as prey for other organisms. Temporarily active geophiles live in the soil in a stable manner for a large part of their life (i.e. for one or more development stages, and emerge from the soil as adults). Most of these organisms are insects, such as Neuroptera, Diptera, Coleoptera and Lepidoptera. Organisms having a "pupae" stage in their life cycle, play a minor role in the soil during these phases, while the "larvae" stage is much more important for the ecology of soil, especially when the population density is high (Jeffery et al., 2010; Wallwork, 1970). Most larvae can act as both detritivores and predators. Periodical geophiles spend a part of their life cycle in the soil, generally as larvae, but throughout their lives they occasionally go back to the soil to perform various activities, such as hunting, laying eggs or to escape dangers. Several Coleoptera groups (e.g. carabids, scarabeids and cicindelids) spend their larval stage in the litter or in the upper layers of mineral soil, and when adults, use soil as a food source, a refuge and for other purposes (Jeffery et al., 2010; Wallwork, 1970). Geobionts are organisms that are very well adapted to life in soil and cannot leave this environment, even temporarily, having characteristics that prevent survival outside of the soil environment such as their lack of protection from desiccation and temperature fluctuations, as well as the sensory organs necessary to survive above ground, for finding food and for avoiding predators. Several species of myriapods, isopods, Acari, molluscs and the majority of Collembola, Diplura and Protura, belong to this group (Jeffery et al., 2010). These different types of relationships between soil organisms and the soil environment determine a differentiated level of vulnerability among various groups, as a consequence of any possible impact on soil environment. For instance, if soil contamination occurs, any impact will be highest on geobionts (because they cannot leave the soil and must spend all their life there) and lowest on temporarily inactive geophiles.

The Role of Soil Fauna in Wood Ecosystems

Soil biota play an essential role in soil functions as they are involved in processes such as the decomposition of organic matter, the formation of humus and the nutrient cycling of many elements (nitrogen, sulphur and carbon). Moreover, edaphic fauna affect the porosity and aeration of, as well as the infiltration and distribution of organic matter within, soil horizons. The ecosystem services provided by soil fauna are one of the most powerful arguments for the conservation of edaphic biodiversity. Decomposition of organic matter by soil organisms is crucial for the functioning of an ecosystem because of its substantial role in providing ecosystem services for plant growth and primary productivity (Maharning et al., 2008). The criteria used to assemble species in groups can be summarized as life-history tactics, microhabitat, principal food, feeding mode and ecophysiology (Brussaard, 1998).

The profile of a mature and undisturbed soil can be divided into a number of layers, whose chemico-physical properties are determined by vegetation cover, by geographical location and by climate; this stratification is reflected in the edaphic population and as a result in its vertical distribution. Usually two groups of organisms are recognised in the soil: the euèdaphons, which include organisms that live in the strip of mineral soil, and the hemiedaphons, found in the strip of organic soil; to these can be added the epiedaphons (or epigeons), consisting of organisms that live in the surface of the soil, and *hyperedaphons* that extend to the herbaceous layer. The moisture content and the pH of the soil have the greatest influence on the distribution of the hemi- and euedaphon fauna, even though the characteristics of the litter, the porosity of the soil and numerous other factors are also important in determining the vertical distributions of edaphons. Humidity has often been used to establish further subdivisions within the hemiedaphons; one example is the classification in which hemiedaphon springtails were subdivided into three categories: hydrophilic (living on the surface of free water), mesophilic (living in damp organic litter) and xerophilic (living in more exposed dry areas, such as in lichens, mosses and tree bark). In reality this type of stratigraphical classification is difficult to apply since soil organisms migrate on both a daily and a seasonal basis; in fact, many species, for example mites, springtails and isopods, can move towards the surface over distances ranging from a few millimetres to a few centimetres. An interesting case is that of Symphyla that can go down as much as 40-50 cm. Nematodes are usually most abundant in the first 10 cm of soil, but their distribution can vary greatly depending on the type of vegetation cover and conditions of dampness. In soil with abundant vegetation cover and roots, nematodes are concentrated in the upper part of the ground which, being rich in exudates and decomposing organic matter, maintain a numerous population of their potential prey. In soils with sparse vegetation cover, the high frequency of dry conditions on the surface causes nematodes to be more common at a depth of 5-10 cm whereas, in cultivated land a uniform distribution of the population can be observed down to a depth of about 20 cm, attributable to the mixing effects of ploughing. Many groups belonging to soil fauna experience seasonal fluctuations in their density in response to the variations in the temperature and humidity of the soil. In a recent study Galli et al. (2012) observed that Protura abundance registered two peaks in March-April and June, while numbers were very low in August and September in a *Quercus suber* wood.

In their turn soil animals differ in their kinds of effects and temporal and spatial scales of influence. We can consider (from Brussaard, 1998):

- Organisms that live in association with a living plant, either beneficially or detrimentally affecting plant growth. These organisms, such as mycorrhizas and many N-fixing bacteria, may live in symbiosis with the root or they may consume root materials (diseases and pests). They are called `root biota'.
- Microflora and micro-/mesofauna acting as regulators of the numbers and activities of microorganisms and microbial feeders. They occur in the `bulk' soil, in the rhizosphere where root-derived materials form their substrate, and in `hot spots' where concentrations of dead organic matter form their substrate. This group is called `the decomposers', to which the meso-/macrofauna that in addition comminute litter entering the soil, without physically reworking the mineral part of the soil, also belong.
- Meso-/macrofauna that create microhabitats for other soil biota by reworking the soil, the `ecosystem engineers'.

Due to the absence of light, which makes photosynthesis unfeasible, among the organisms populating the soil we find very few real phytophages, unless we extend the definition of soil animals to include surface organisms, or if we consider that pests also include fungivorous microarthropods (Wallwork, 1970). The activity of animals, among them typically protozoa, nematodes, rotifers, certain springtails and mites, which feed on microflora, consisting of bacteria, actinomycetes and fungi (both hyphae and spores), is of crucial importance both for regulating the density and for the diffusion of these micro-organisms. For example, through their faeces, springtails, which feed on fungi, can spread fungal spores that are still viable to areas as far as a few metres away from their point of origin. Also, feeding on microflora by the fauna at the same time controls specific plant pathogens.

The detritus food chain takes on an essential role within the soil, as it becomes the basis of the hypogean food web; in fact, many organisms such as isopods, certain myriapods, earthworms, springtails, many species of mites, and the larvae and adults of many insects feed on the vegetable and animal detritus that is deposited on the soil. Soil fauna perform a mainly mechanical action, whereas chemical degradation is essentially performed by fungi and bacteria, both the free and intestinal symbionts of other organisms; furthermore during digestion, organic substances are enriched by enzymes that are dispersed in the soil along with the faeces, contributing to humification. Nutrients are released through faunal-microfloral interaction. Microflora are able to extract inorganic ions both from organic substrates and from the soil solution, resulting in a higher concentration of nutrients in their biomass than in the external substrate (Marshall, 2000). The C:N of bacteria is very close to that of the fauna, hence faunal grazing on the microflora results in a net release of nitrogen (Marshall, 2000). Nutrients temporarily immobilized by bacteria and fungi are eventually released by faunal feeding and finally by the death of the microflora and fauna themselves (Marshall, 2000).

The decomposition rate is generally positively correlated with faunal biomass, and a diverse fauna is required for interacting with microorganisms to enhance microbial diversity (Marshall, 2000). Earthworms are among the most important organisms in many of the soils of the world. According to an ecological and functional classification (Bouché, 1972) it is possible to identify three groups of earthworms that differ in size, burrowing capability, type of food and habitat: epigeic, endogeic and anecic. The epigeic, whose sizes range from 0.5 to 5 cm, have a red coloured body, are poorly adapted to burrowing and have a good tolerance to low pH values. They inhabit the superficial organic layers where they feed on litter. The endogeic, whose sizes range from 1 to 8 cm, have a non-pigmented body; exposure to soil with pH values below 5 restricts their activity. They live in the first few centimetres of mineral soil, feeding on humic compost and dead roots, they are capable burrowers and they make tunnels that extend mainly horizontally. Earthworms defined as anecic, whose sizes generally are in excess of 5 cm and have a reddish-brown body, are excellent burrowers, making vertical tunnels that can reach several metres in depth; they tend to avoid soil that is asphyxiated or lacking in moisture. They live in the mineral layer of the soil but rise to the surface, mainly at night, to feed on litter. The activity of earthworms produces a significant effect, not just on the structure, but also on the chemical composition of the soil, since a large part of the organic matter ingested by earthworms is returned to the soil in a form easily used by plants. While they are feeding, earthworms also ingest large quantities of mineral substances (minimally so in the case of the epigeic), that are then mixed with the organic matter ingested and, after having been cemented with a little mucous protein, are expelled in piles called worm casts. In addition to being rich in nitrogen and other nutritive substances such as calcium, magnesium and potassium, worm casts also contain a large quantity of non-digested bacteria which proliferate easily in this sub-stratum and contribute to the humification and mineralisation of organic matter (Zanella et al., 2001). Vermicasts have more favourable physico-chemical properties, an increased microbial population, and enzyme activities and nutrient mineralization that support plant growth and yield (Parthasarathi et al., 1999). Many studies have reported increased microbial activities during the passage of food through the gut in earthworms (Burges and Raw, 1967; Parthasarathi et al., 1999) and higher numbers of fungi, bacteria and actinomycetes in vermicasts (Burges and Raw, 1967).

Spherical soil concretions composed of calcite crystals have been found in earthworm casts and soils containing earthworms. These calcite granules range from single calcite crystals to aggregations up to 2.5 mm in diameter and are produced by the calciferous glands of all species of earthworms of the Lumbricidae family (Wiecek and Messenger, 1972; Canti and Piearce, 2003; Briones et al., 2008a; Gago-Duport et al., 2008; Lee et al., 2008). The calcium carbonate is present predominantly as calcite but also as aragonite, valerite and amorphous calcium carbonate (Gago-Duport et al., 2008). The function that calcium carbonate granule secretion serves for the earthworm is unknown with suggestions linking it to excretion of excess Ca, neutralisation of gut pH and regulation of CO_2 (Lambkin et al., 2011). Coleman et al. (2004) and Briones et al. (2008b) indicate that the calciferous glands of earthworms could provide a mechanism for regulating CO_2 in their blood and tissues.

It is important to note that often worm casts are not released in the same layer in which the earthworm fed; in effect the components of the anecic group live at depth and release their mineral-rich faeces on the surface. In contrast the endogeic live at the surface and release their faeces, once again rich in organic matter, at lower depths. These forms of behaviour together with the direct action due to the burrowing, ensure that the soil is mixed, thereby increasing its fertility. The burrowing of the earthworms is also essential for increasing the aeration of the soil and for improving the circulation of water since their tunnels increase the porosity of the ground by 20-30% (Zanella et al., 2001), enabling those organisms that are not good burrowers to move around easily even at lower depths of the soil. Other than with their worm casts, earthworms contribute to the increase in the amount of nitrogen present in the ground through the excretion of ammonia and urea, forms that are directly useable by plants; furthermore a sizeable quantity of nitrogen is returned to the soil on the death of animals, which have a 72% protein content (Dindal, 1990).

The soil fauna, in particular molluscs and earthworms, also have an effect on the soil through the secretion of cutaneous mucous, that has a cementing effect on the particles in the ground, assisting the stability and structure of the soil and making it less vulnerable to processes of erosion. The mucous secretions, the faeces (especially those of earthworms) and the bodies themselves of the animals (when they die) influence in large measure the concentration of nutrients present in the soil particularly potassium, phosphorous and nitrogen, reducing the C:N ratio of the litter and facilitating decomposition. Siddiky et al. (2012) demonstrated that collembola populations can play a vital role in maintaining high levels of soil aggregation, including when they co-occur with another agent of soil aggregation, arbuscular mycorrhizal fungi.

Yang and Chen (2009) showed that the soil fauna community played a more substantial role in leaf litter decomposition in the tropical rain forest than in broad-leaf and secondary forests by significantly accelerating the decomposition rate.

The authors suggested that soil fauna play a critical role in the decomposition of leaf litter in tropical rain forest by modifying the initial low N availability and high C:N ratio, which limit microbial activity. The data reported by the authors indicated that the soil fauna decrease the C concentration and enhance the N concentration in the rain forest leaf litter, where the initial high C:N ratio is substantially different from initial conditions in secondary and broad-leaf forests.

A very interesting environment present in the soil is the rhizosphere. The term rhizosphere is used in its broadest sense to refer to the portion of soil surrounding roots in which the micro-organisms are influenced by their presence; its extension varies greatly but in general it is considered to be the cylinder of soil used by the root hairs and in which they emit exudates (Killham, 1994).

It can be distinguished from the majority of the soil on the basis of its chemical, physical and biological characteristics. As they penetrate the ground, the roots act on clay minerals and the particles of soil surrounding them; this leads to the formation of a surrounding area in which the water pathway, and the movement of nutrients and microflora is more heavily channelled than in the rest of the soil. For the same reasons, the organic matter released by the roots accumulates close to them. The chemical nature of the rhizosphere is significantly different from the rest of the soil; this results in large part from the release of carbon and the selective capturing by the roots of ions in solution in the groundwater.

The plants act as carbon pumps fixing what is available in the atmosphere in the root exudates which are quickly captured by soil organisms; for this reason the level of carbon available around the roots is never very high. The selective absorption of ions instead, causes the depletion of some of them in the rhizosphere, while others, not absorbed by the roots, tend to accumulate.

There can be a very close relationship between the roots and the microflora leading to (1) bacteria or fungi becoming an integral part of the roots such as in mycorrhizal symbiosis and (2) in the association of bacteria and legumes.

The peculiar characteristics of the rhizosphere are also reflected in a selectivity of the animal element. Protozoa and nematodes are the main consumers of microorganisms developing near living and dead roots and in hotspots in the soil matrix; rhizosphere: nonrhizosphere soil ratios are in the range of 1.3-35 for protozoan numbers and 3-70 for nematode numbers (Brussaard, 1998).

The interaction between soil animals and the roots of plants can take a variety of forms that lead to benefits or that repress the growth of the plants, and often involve interactions with the microbial populations of the soil (Lammed, 2012). Root-feeding nematodes interact with mycorrhizas. The nematodes may reduce the growth stimulated by the mycorrhizas or the mycorrhizas may compensate in part for the growth reduction caused by the nematodes (Brussaard, 1998).

The dispersion of the inocula of mycorrhizal fungi by soil animals can have beneficial effects for the plants; this dispersion is particularly favoured by burrowing organisms belonging to the mega- and mesofauna. The hyphae of mycorrhizal fungi may make up a significant proportion of the total microbial biomass in some soils and can become one of the most important sources of food for fungus grazing animals such as springtails.

Numerous soil animals feed directly on the roots of plants; among them are a large number of species of springtails and myriapods. It is still not entirely clear how much of the damage inflicted on plants can be attributed to the direct action caused by the grazing of the roots and how much to the subsequent vulnerability of the roots to pathogens in the soil, especially fungi. Root herbivory influences individual plant performance and higher-level processes both directly and indirectly (Brussaard, 1998): directly by the selective feeding on seeds and seedlings, thereby preventing the establishment or abundance of certain plant species and indirectly by interfering with the carbon and nutrient acquisition and allocation of plants in interaction with above-ground herbivores.

DIVERSITY OF SOIL FAUNA IN DIFFERENT WOODLANDS

The abundance, diversity, composition and activity of species of the soil community can be affected by plant species, plant diversity and composition, as well as by animal grazing (Maharning et al., 2008). The two factors that are most likely to influence the soil community are the nutrient resources available, and the diversity of microhabitats (Maharning et al., 2008). Resource type is determined by the tissue chemistry of the plant species and the nutrient content of excreted wastes from grazing by consumers (Maharning et al., 2008). The microhabitat is directly affected by the physical and chemical properties of the system as well as the amount of nutrient input (Bardgett et al., 1998).

Various studies have described the structure of soil invertebrate communities in relation to forest diversity, dynamics and management (Bird et al., 2000; Theenhaus and Schaefer, 1995). Several studies considering the effects of silvicultural practices on soil fauna found important impacts on forest soil fertility/productivity and in the terrestrial food chain (Moore et al., 2002). Harvesting directly affects the mineralization of essential plant nutrients, and the conservation of these nutrients within the soil system through the reduction and redistribution of organic matter, compaction, changes in plant cover, and modification of microclimate, all of which affect the distribution, composition and activity of the soil biological communities (Marshall, 2000).

It is generally accepted that the removal of trees by clear-cutting, or other methods, has a significant effect on the invertebrate fauna of the forest floor (Heliovaara and Vaisanen, 1984; Hoekstra et al., 1995). For many years dead wood has been an important element in the forestry management of many European countries. Studies of forest ecology have for some time highlighted the importance of the presence of "dead wood" in forests (referred to by the acronym CWD = Coarse Woody Debris). Dead wood is found in whole or partially decomposed trunks and stumps with a diameter greater than 8-10 cm, which have become the habitat of plants, animals and insects as well as being a source of nutrients for the formation of soil. Moreover CWD offers sheltered microhabitats and may serve for food and as a site for breeding (Harmon et al., 1986). The decomposition of dead wood takes place through the activity of microflora, and soil fauna organisms, and therefore derives directly from the population underneath, and of creatures that colonise the wood. The presence of dead wood in woodlands produces numerous positive effects among them that it helps improve the stability of the ground, slowing down erosion and run off, it adds organic substances to the soil, it contributes to its evolution and becomes an important part of the biogeochemical process. It also provides a habitat for decomposer organisms and, by retaining moisture, helps them survive through the most difficult periods. In addition it provides a useful substrate for the germination of seeds, encouraging the natural renewal of the woods, it becomes a den, a refuge or source of nutrients for a multitude of creatures, from mammals to insects, from worms to amphibians (about 1/5 of all animal species that live in a forest are linked in some way to dead wood).

Dead wood is broken down by the activity of xylobiontic organisms, that feed directly on dead wood or bark and of organisms which, during one or more phases of their biological cycle, depend on the presence of dead wood. This functional grouping includes protozoa, nematodes, insects, myriapods, etc. So included among the xylobiontic organisms are xylophage organisms which feed directly on the dead wood and tenant organisms which colonise the tunnels bored out by xylophages to deposit their eggs, pass the winter, find protection, etc.

The decomposition of dead wood can require differing periods of time, sometimes very long, depending on the species of tree which, due to its nitrogen content, the proportion of the cellulose to the lignin polymers, the consequent C:N ratio, the presence of organic acids, tannins, phenols and other complex substances, may be more or less easily biodegradable and appetising to the detritivores and decomposers. The climate, particularly the temperature and damp, plays a leading role in the times required for decomposition, either favouring or inhibiting the activity of the microflora and the xylobiontic organisms.

The process of transforming dead wood leads to the formation of minerals and partially decomposed organic substances and to neosynthesis, which supplements the amount of humus in the soil's O horizon. Hence also involved in this process are organisms which break down the organic matter and which participate in the normal recycling of matter such as heterotrophic bacteria, actinomycetes and fungi, which use cellulose, lignin, and other kinds of molecules to obtain energy. The high numeric concentration of these organisms encourages the development of microphage organisms such as nematodes, protozoa and springtails, which find dead wood to be an excellent habitat both as a food source, and for their reproduction processes. In their turn, these organisms attract numerous larger creatures such as dermaptera, centipedes, beetles, pseudoscorpions, araneidae and small scorpions, which actively prey on the micro- and mesofauna that collect in the wood. A recent study has shown how all the higher taxa of the soil macrofauna occurred in higher numbers at sites close to CWD than at sites far from it. Differences were most evident for the zoophagous Chilopoda and Araneide, the saprophagous Isopoda, and Coleoptera-adults and Coleoptera-larvae (Jabin et al., 2004). The authors found that CWD not only increases species diversity but also functional diversity.

FACTORS AFFECTING THE LOSS OF SOIL FAUNA DIVERSITY IN WOODLANDS

Humans have been the cause of a major global reduction in biodiversity. The Convention on Biological Diversity (CBD) was the first global agreement aimed at conservation and the sustainable use of biological diversity (Secretariat of the Convention on Biological Diversity, 2000). The CBD lies at the heart of biodiversity conservation initiatives. It offers opportunities to address global issues at a national level through locally grown solutions and measures. One important requirement is the development of National Biodiversity Strategies and Action Plans mainstreaming them into relevant sectors and programmes, a principal means for implementation of the Convention at the national level (United Nations 1992). The recent Conference of the Parties of the CBD (May 2008, Bonn) demonstrated that the need for action to protect biodiversity is unanimously acknowledged. Biodiversity conservation is essential both for ethical reasons and especially for the ecosystem services that the complex of living organisms provides for current and future generations. These ecosystem services are essential for the functioning of our planet (Menta, 2012).

The changes in biodiversity alter ecosystem processes and change their resilience to environmental change. It is estimated that human activities increased the rates of extinction 100-1000 times (Lawton and May, 1995). In the absence of major changes in policy and human behaviour our effect on the environment will continue to alter biodiversity.

Land use is considered to be the main element of global change for the near future, and in this regard, land use change is projected to have the largest impact on biodiversity by the year 2100. In a review on changing biodiversity, Chapin et al. (2000) consider that land use will be the main cause of change in biodiversity for tropical, Mediterranean and grassland ecosystems.

As far as soil biodiversity is concerned, forests, tropical or temperate, generally represent the biomes with the largest soil biodiversity. Consequently any land use change resulting in the removal of perennial tree vegetation will produce a reduction of soil biodiversity. In some cases pasture or perennial grasslands succeed forests while in others, arable land replaces formerly wooded areas. The change in soil biodiversity will therefore be influenced by the subsequent use of land following the forest (Arredondo et al., in press). On the other hand, reduction of soil biodiversity as a result of urbanization can be even more severe. The urbanization process leads to the conversion of indigenous habitat to various forms of anthropogenic land use, the fragmentation and isolation of areas of indigenous habitat, and to an increase in local human population density. The urbanization process has been identified, for example, as one of the leading causes of decline in soil arthropod diversity and abundance in some areas (Menta, 2012). The result of these soil biodiversity reductions are artificial ecosystems that require constant human intervention and extra running costs, whereas natural ecosystems are regulated by plant, animal and soil organism communities through flows of energy and nutrients, a form of control progressively being lost with agricultural intensification.

Furthermore differences in agricultural and forestry production systems, such as integrated, organic or conventional systems, have been demonstrated to affect soil fauna in terms of numbers and composition (Hansen et al., 2001; Cortet et al., 2002). Taking the example of communities of soil microarthropods, observations on the impacts of different forms of agricultural and forest management on these communities showed that the high input of intensively managed systems tends to promote low diversity while lower input systems conserve diversity (Menta, 2012).

Sousa et al. (2004) reported a study regarding the effects of land-use on Collembola diversity patterns in a Mediterranean landscape. In this study Collembola communities were sampled along a gradient of soil-use intensification in a typical Mediterranean landscape dominated by cork-oak. This gradient ranged from a land-use unit dominated by closed oak forest with minimum intervention to a unit composed entirely of a monoculture crop submitted to frequent anthropogenic disturbances, and in between others with managed woodland or dominated by open cork-oak areas and pastures. Abundance, diversity and species richness decreased along the gradient, with the agricultural site presenting an impoverished community while the first site (oak forest with minimum intervention) was where most species and the highest density were identified (Sousa et al., 2004).

Déchêne and Buddle (2009) in a study in connection with the effects of experimental forest harvesting on oribatid mite biodiversity, suggested that oribatid mite assemblages were affected in various ways by different harvesting practices in eastern boreal forests, but that small-scale regional factors may be even more important in shaping oribatid assemblages.

Finally, in biodiversity recovery processes using forest restoration, it must be taken into account that, having stimulated natural regeneration processes that establish forest species, it is necessary to manage and direct succession processes toward the desired objectives. It is important to promote the continued development of the vegetation to conserve soil, nutrients, and organic resources; to restore fully functional hydrological processes, nutrient cycling and energy flow processes; and to create self-repairing landscapes that provide the goods and services necessary for biophysical, biochemical and socioeconomic sustainability.

Different stages of forest soil degradation require management actions that focus on different processes. Severely degraded sites require early repair of hydrological, nutrient cycling, and energy capture and transfer processes. As the forest vegetation increases in biomass and stature, it reduces the abiotic limitations of the site by improving soil and micro-environmental conditions. Directing natural processes toward land use goals requires an understanding of the processes driving succession. The rate and direction of succession is influenced by the availability of species, the availability of suitable sites, and by the performance of differential species (Mansourian, 2005).

Microarthropod Communities in Different Italian Forest Soils – Some Case Studies

In forest ecosystems soil and litter arthropod communities play a major role in the decomposition of fresh organic matter and in the formation of the humus profile (Ponge et al., 1986; Bird et al., 2000). Microarthropods, particularly Oribatid mites and Collembola are important members of the detrital system (Seastedt and Crossley, 1981). They participate in the fragmentation of plant detritus and stimulate the activity of bacterial and fungal colonies. The soil arthropod community is often considered to be a useful bioindicator of forest condition and change (Hole, 1981) as soil organic matter dynamics and nutrient mineralization are key processes in ecosystem functioning (Bardgett et al., 1998). Despite this information, the relationship of animal communities inhabiting soil below different understory plants, in different moisture regimes, dead organic matter and other patches on the forest floor to the “main” community of a specific forest has not yet been well studied (Rusek, 2001).

In this part of the chapter we show and underline differences in microarthropod communities associated with different lands and, in particular, different woodlands.

Different datasets, from about ten years of research, were analyzed in order to compare the effect of different types of vegetation on the microarthropod community. For each site the QBS-ar values (an index of biological quality, Parisi et al., 2005), the number of taxonomic groups and the presence/absence of each group were recorded. The study areas were:

- Agricultural lands that include areas cultivated with different crops (wheat, maize and barley) from different farms almost all located in Northern Italy;
- Spruce woods in the Vajont River valley characterized by *Picea abies*. Spruce forests grow in cold weather where there is an abundance of water, creating a moist and shaded environment in which there is an accumulation of ectohumus (raw humus);
- Permanent grasslands between the ages of 5 and 150 years, located in the “Taro River Regional Park” (Northern Italy). Permanent grasslands are meadow that is maintained perpetually without reseeding;
- Chestnut woods situated in the “Boschi di Carrega” Regional Park (Northern Italy). This type of forest is associated with acid soil and wet weather;
- Beech woods including different sites spread all over the Italian Apennine, characterized by the presence of *Fagus sylvatica*. Beech forests grow in the specific climatic conditions related to the altitudinal range typical of this species; moreover the weather directly affects the pedological conditions of beech forests. It is characterized by the accumulation of acid humus during winter and mild humus, which preserves and enhances soil fertility, during summer;
- Oak woods characterized by the presence of *Quercus cerris*, *Quercus petraea* and *Quercus pubescens*. Like all oak forests, they grow in temperate climates with an abundance of water, creating an environment which is shaded during summer and bright during winter and in which there is an accumulation of mild humus.

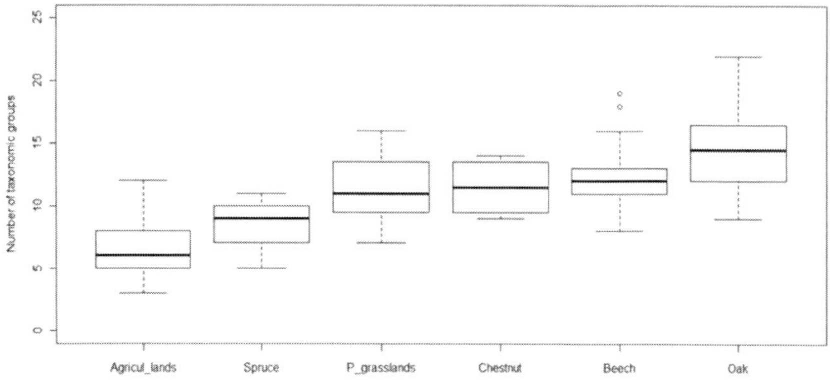

Figure 1. Number of taxonomic groups recorded in different areas.

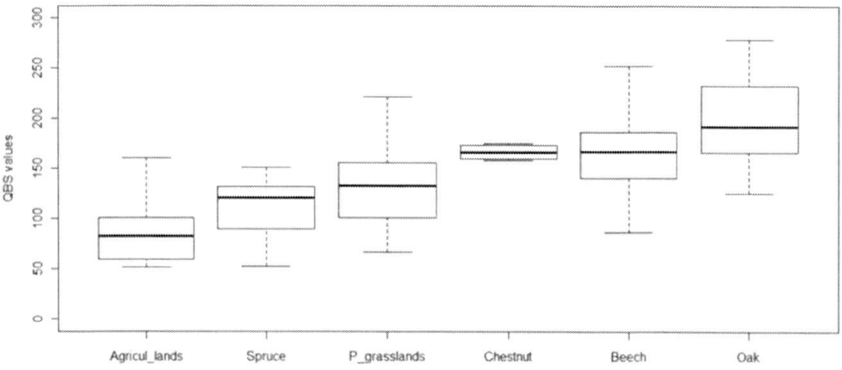

Figure 2. QBS-ar values recorded in different areas.

Several groups of microarthropods were found in each area with the lowest score being associated with agricultural lands and the highest, with oak woods (Figure 1).

In particular we recorded 6 taxonomic groups (median value) for agricultural lands, 9 for spruce woods, 11 for permanent grasslands, less than 12 for chestnut woods, 12 for beech woods and more than 14 for oak woods. These results showed how the higher numbers of taxonomic groups were associated with forests (except for spruce woods), and the lower with agricultural lands. Results associated with agricultural lands are in agreement with the trends usually observed in these areas and are related to a significantly lower number of groups generally associated with stable soil conditions, such as pauropods, symphylans, proturans and diplurans. This could be related to the soil management typical of agricultural lands. We recorded lower values for spruce woods than for permanent grasslands. Meadows are characterized by a higher biodiversity of plants while, probably, spruce woods were not suitable for the development of a well diversified microarthropod community. The C:N ratio of grassland litter is usually low (<20:1) and is reflected in rapid decomposition. By contrast much coniferous litter has a C:N ratio exceeding 60:1 and its decomposition may take several years (Curry, 1994). Moreover, undisturbed natural grasslands are characterized by a relatively higher organic matter content and a more even distribution with depth in the profile when compared with adjacent deciduous land on comparable soil. This is due to the more rapid turnover of the relatively short-lived grasses, and the greater proportion of root to soil organic matter in grassland (Curry, 1994). Finally the highest values were found in woodlands. In particular, the results associated with oak woods suggest a well diversified edaphic fauna while the lower values associated with beech woods are probably due to the lower palatability of beech litter and to the particular climatic conditions typical of the altitudinal range of beech forests.

The same analyses were carried out for the QBS-ar values (Figure 2). This index is based on the following concept: the higher the soil quality, the higher will be the number of microarthropod groups which are morphologically well adapted to soil habitat. Soil organisms are separated into biological forms in accordance with their morphological adaptation to soil environments; each of these forms is associated with a score called the EMI (eco-morphological index), which ranges from 1 to 20 in proportion to the degree of adaptation. The QBS-ar index value is obtained from the sum of the EMI of all collected groups (for more details see Parisi et al., 2005 and Menta et al., 2011).

The QBS-ar analysis revealed the same trend as that recorded in the analysis of the numbers of taxonomic groups, with the higher values being associated with oak woods and the lowest with agricultural lands. In particular we recorded a QBS-ar value of 83 (median value) for agricultural lands, 121 for spruce woods, about 133 for permanent grasslands, about 167 for chestnut woods, 167 for beech woods and about 193 for oak woods. Once again, the higher values were associated, with forests (with the exception of spruce woods) emphasizing a better soil quality for these areas.

An analysis of the density of organisms showed similar trends with the higher values associated with permanent grasslands (Figures 3-8). In particular springtails and mites reached the highest values in this area, while other taxonomic groups were poorly represented. Agricultural lands showed the same trend with the higher values associated with mites. In general mites and springtails were the most representative groups in all the areas. Forest areas also showed higher values for those taxonomic groups well adapted to soil habitat, such as symphylans, proturans and diplurans which are poorly represented in agricultural lands and permanent grasslands. In general in forest areas we recorded well distributed microarthropod communities, in particular for oak and chestnut woods, while spruce and beech woods displayed a predominance of mites and springtails. The latter, for example, showed high densities of mites with levels as high as 76%.

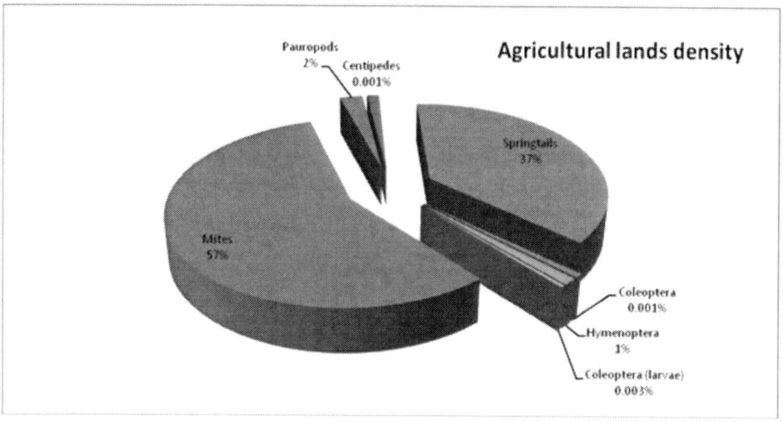

Figure 3. Densities of organisms recorded in agricultural lands.

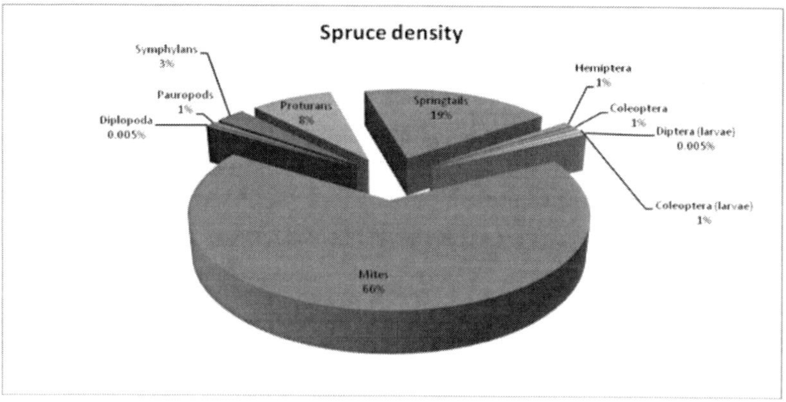

Figure 4. Densities of organisms recorded in spruce woods.

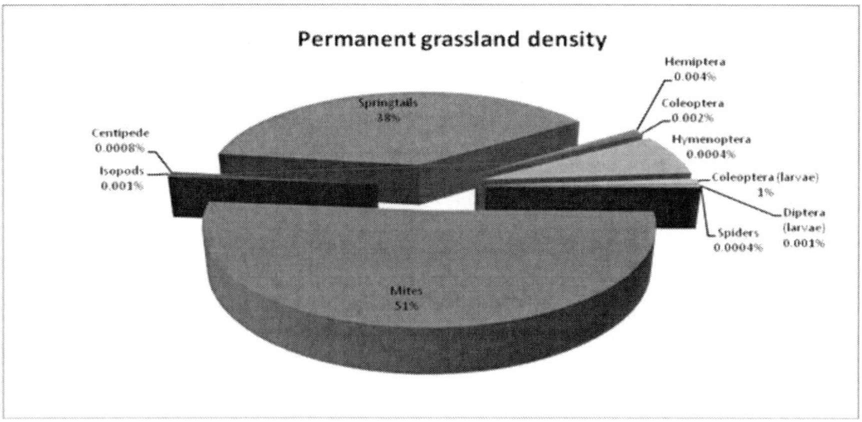

Figure 5. Densities of organisms recorded in permanent grasslands.

Finally, in chestnut woods only, we recorded the presence of palpigrades, rare tropical and warm-temperate arachnids; they are minute, whitish, eyeless animals with an elongated body that terminates in a slender, multisegmented flagellum set with setae.

Finally we want to show how the number of groups and the densities of organisms change between forests and extrasylvatic vegetation (Leoni, 2008 data unpublished). This study was carried out in the "Guadine Pradaccio" Natural Reserve located in the Northern Apennine. We will focus on the results associated with beech and extrasylvatic vegetation sampling (Figure 9).

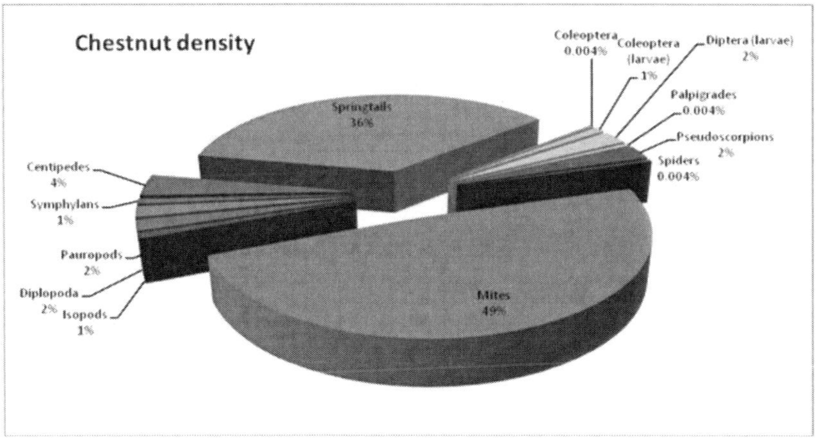

Figure 6. Densities of organisms recorded in chestnut woods.

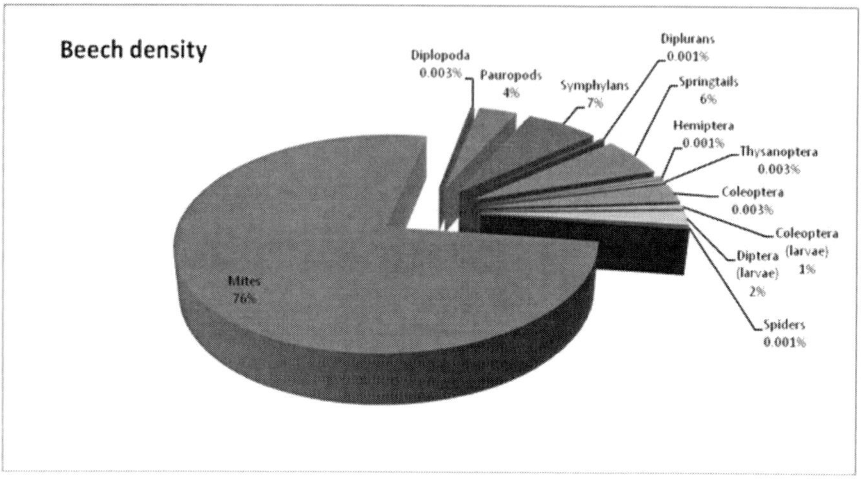

Figure 7. Densities of organisms recorded in beech woods.

An analysis of the number of taxonomic groups and the density of organisms showed a well diversified microarthropod community in beech woods which was in agreement with previous results. In fact the litter of a beech wood is not only a great source of plant debris and an excellent substrate for the growth of mushrooms, on which microarthropods feed, but also plays a role in the regulating of soil moisture (Bernetti, 1995); this is one of the most important parameters that determine the composition of edaphic fauna (Hagvar, 1982; Ponge, 1993; Kuznestova, 2002). Instead brachypodietum was found to occupy an intermediate position having a similar number of taxonomic groups to beech woods but with densities similar to vaccinietum. Finally, vaccinietum returned the lowest numbers of taxonomic groups and densities. This is probably due to the lower level of protection from environmental stresses (thermal and water) and the reduced availability of trophic resources in comparison with forest areas.

The same comparison was carried out for the number of species of springtails recorded in the area (Figure 10). Collembolan are among the most abundant groups in the organic layers of forest soils (Petersen and Luxton, 1982) playing an important role in litter decomposition by grazing soil fungi and stimulating nutrient cycling (Bengtsson and Rundgren, 1983; Hassal et al., 1986).

In this case also we recorded the highest number of species in beech woods while the lowest were in vaccinietum. Once again, brachypodietum results between the two.

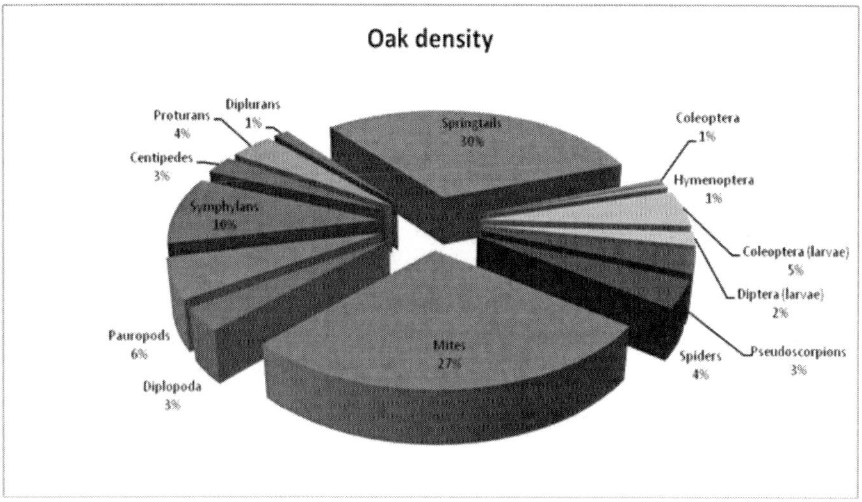

Figure 8. Densities of organisms recorded in oak woods.

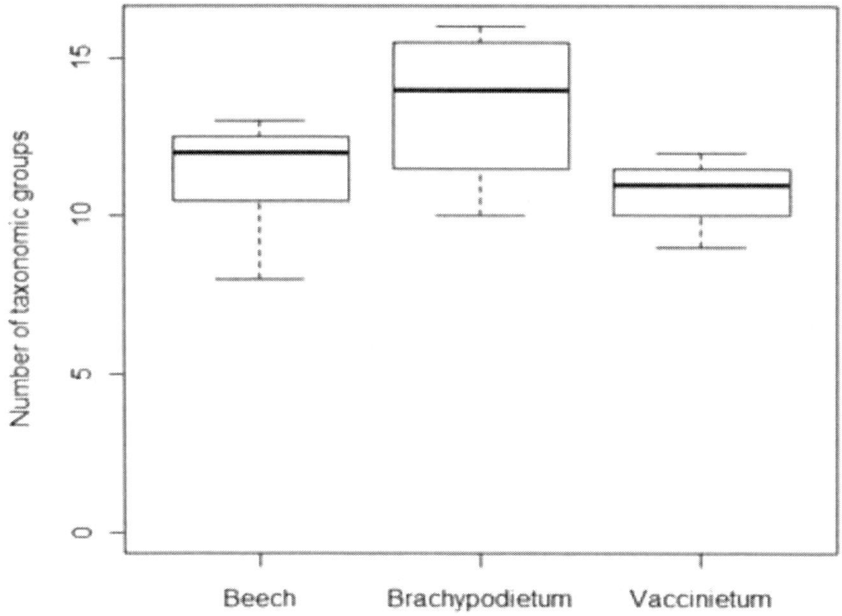

Figure 9. Number of taxonomic groups recorded in beech wood, brachypodietum and vaccinietum.

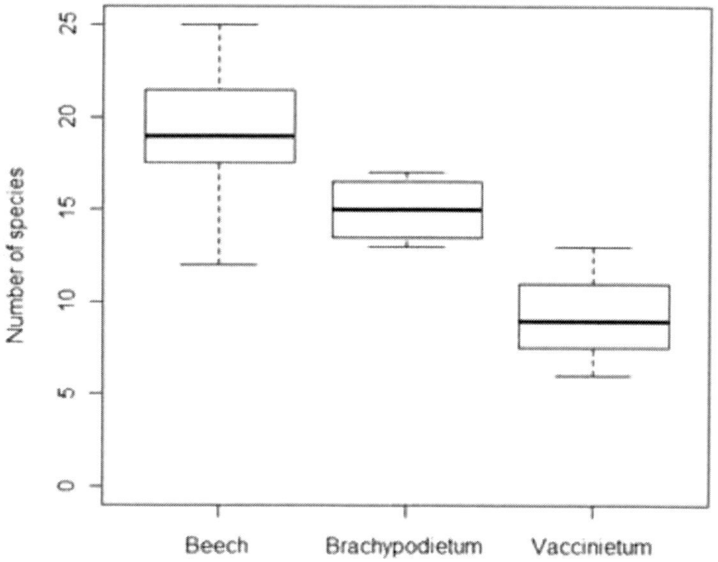

Figure 10. Number of species of collembola recorded in beech wood, brachypodietum and vaccinietum.

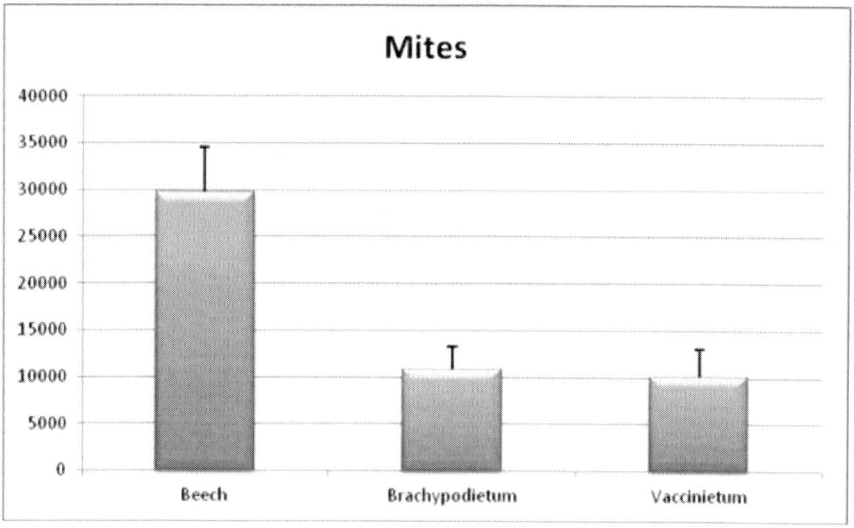

Figure 11. Densities of mites recorded in beech wood, brachypodietum and vaccinietum.

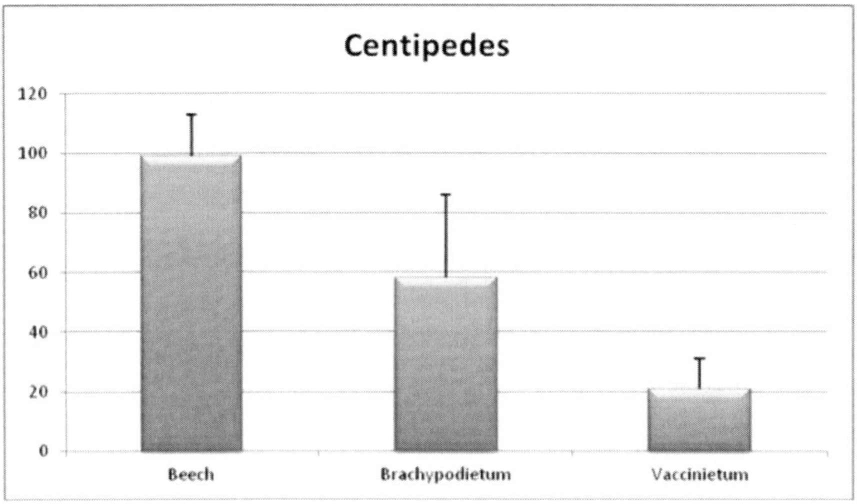

Figure 12. Densities of centipedes recorded in beech wood, brachypodietum and vaccinietum.

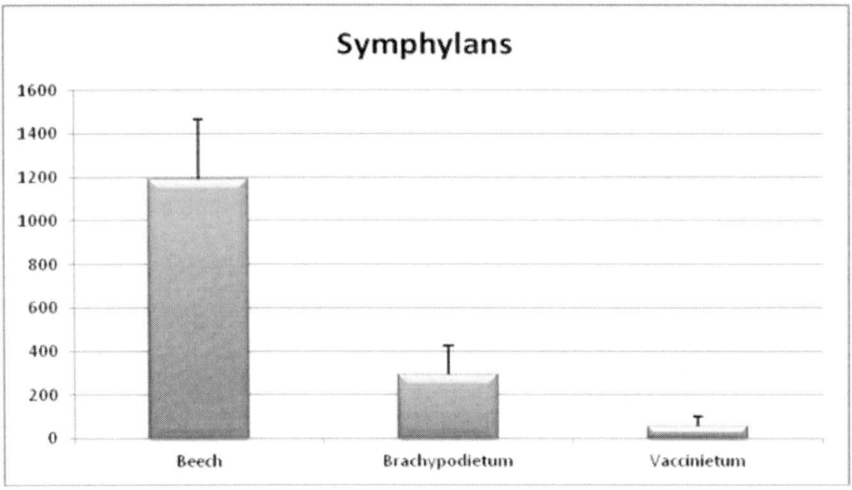

Figure 13. Densities of symphylans in beech wood, brachypodietum and vaccinietum.

Finally all these results were confirmed by the analysis of the densities of certain groups of organisms. Figures 11-17 show how densities decrease from beech wood to brachypodietum and then to vaccinietum.

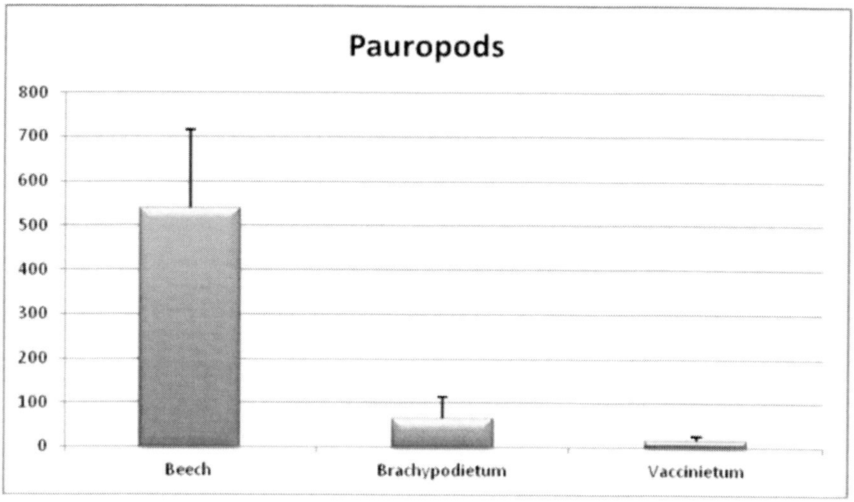

Figure 14. Densities of pauropods in beech wood, brachypodietum and vaccinietum.

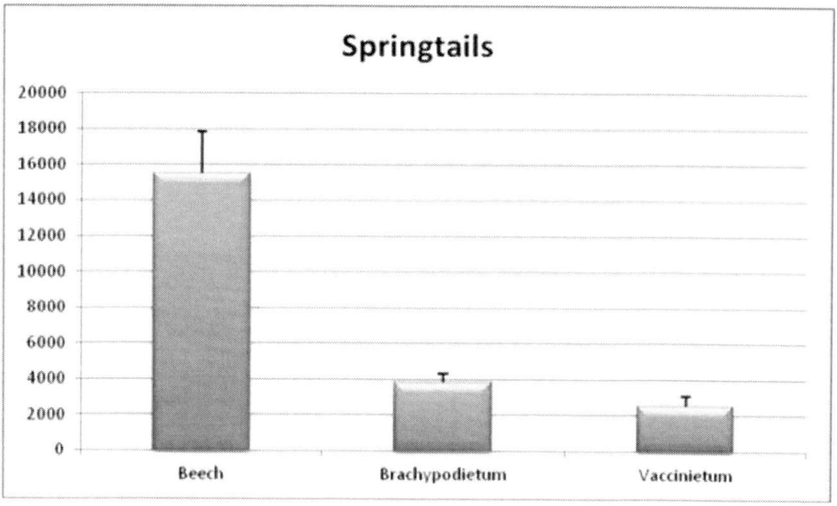

Figure 15. Densities of springtails in beech wood, brachypodietum and vaccinietum.

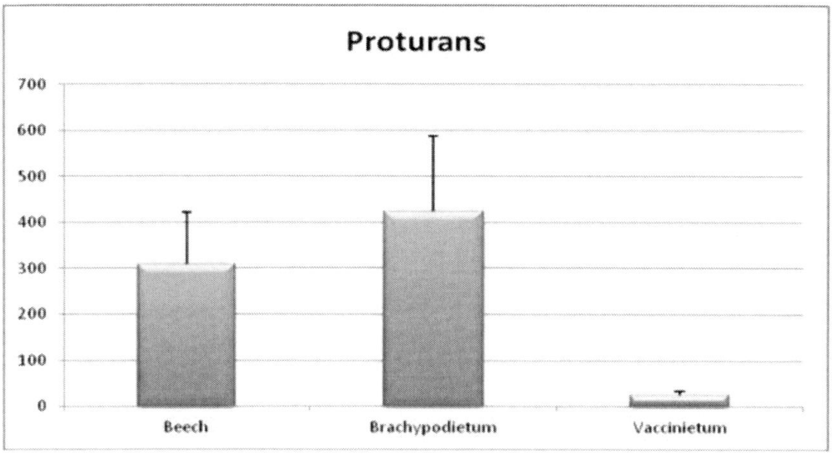

Figure 16. Densities of proturans in beech wood, brachypodietum and vaccinietum.

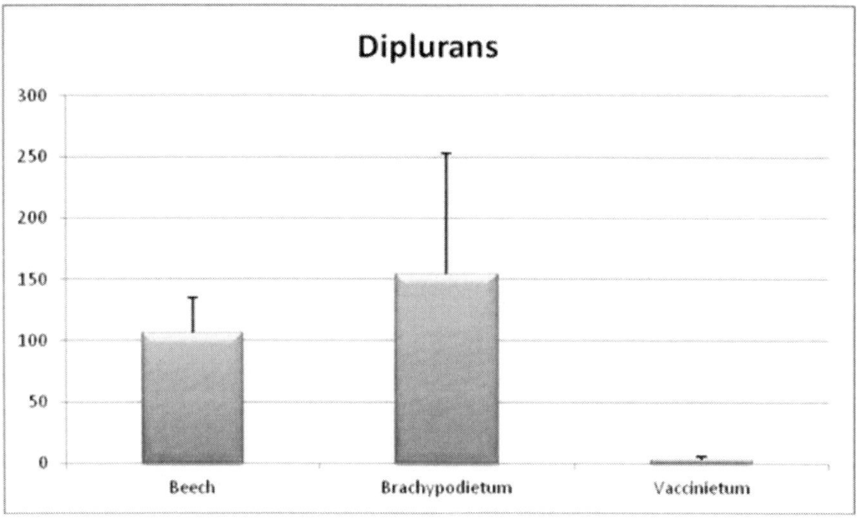

Figure 17. Densities of diplurans in beech wood, brachypodietum and vaccinietum.

Conclusion

Too rarely do we pause to reflect on the fact that soil is the foundation upon which society is sustained and evolves, that it is a vital component of ecological processes and cycles, as well as the basis on which our infrastructure rests. Growing pressures from an ever increasing global population, as well as threats such as climate change and soil erosion, are placing increasing stresses on the ability of soil to sustain its important role in the planet's survival. It has been demonstrated that soil biodiversity increase with the increasing of natural conditions (for example in forest ecosystems) and the decreasing of human impacts (for example impacts related to agricultural lands). Moreover, evidence suggests that while increased use of mono-cultures and intensive agriculture has led to a decline in soil biodiversity in some areas, the precise consequences of this loss are not always clear (Jeffery et al., 2010). Soil is one of the fundamental components for supporting life on Earth. It is the processes that occur within soil, most of which are driven by the life that is found there, which drive ecosystem and global functions and thus help maintain life above ground. Soil performs numerous ecosystem functions and services, ranging from providing the food that we eat to filtering and cleaning the water that we drink. It is used as a platform for building and provides vital products such as antibiotics, as well as containing an archive of our cultural heritage in the form of archeological sites. A more complete knowledge of soil fauna is needed for biodiversity conservation. In fact, the majority of soil organisms are still unknown; despite their important contribute to soil functions such as nutrient cycling, soil formation and water regime control.

Only by knowing soil in all its complexity, while maintaining its functionality and quality through actions aimed at protecting its properties, and acknowledging the importance it assumes in the quality of life worldwide, can we embark on a truly sustainable use of soil perceived as a resource and build a proper Man / Soil relationship to be left to future generations.

References

Arredondo, F.; García-Montero, L.G., Menta, C.; Valverde-Asenjo, I. (in press.). Soil quality indicators for forest management. In: Quantitative Techniques in Participatory Forest Management. Martínez-Falero, E.;

Martín-Fernández, S.; García-Abril, A. (Eds.). *Taylor and Francis Group*, Boca-Ratón.

Bardgett, R.D.; Keiller; S.; Cook, R.; Gilburn, A.S. (1998). Dynamic interactions between soil animals and micro-organisms in upland grassland soils amended with sheep dung: a microcosm experiment. *Soil Biol. Biochem.*, 30(4), 531-539.

Bengtsson, G.; Rundgren, S. (1983). Respiration and growth of a fungus, *Mortierella isabellina*, in response to grazing by *Onychiurus armatus* (Collembola). *Soil Biol. Biochem.*, 15, 469-473.

Bernetti, G. (1995). Selvicoltura speciale. Scienze Forestali ed Ambientali. UTET

Bird, S.; Coulson, R.N.; Crossley, Jr. D.A. (2000). Impacts of sylvicultural practices on soil and litter arthropod diversity in a Texas pine plantation. *For. Ecol. Manag.*, 131, 65-80.

Bouché, M.B. (1972). Lombriciens de france. Ecologie et systematique. *INRA 72-2 Institut National Des Recherches Agriculturelles*, Paris.

Briones, M.J.I.; Lopez, E.; Mendez, J.; Rodríguez, J.B.; Gago-Duport, L. (2008a). Biological control over the formation and storage of amorphous calcium carbonate by earthworms. *Mineral Mag*, 72, 227-231.

Briones, M.J.I.; Ostle, N.J.; Piearce, T.G. (2008b). Stable isotopes reveal that the calciferous gland of earthworms is a CO_2-fixing organ. *Soil. Biol. Biochem.*, 40, 554-557.

Brussaard, L. (1998). Soil fauna, guilds, functional groups and ecosystem processes. *Appl. Soil Ecol.*, 9, 123-135.

Burges, A.; Raw F. (1967). Soil Biology. Academic Press.

Canti, M.G.; Piearce, T.G. (2003). Morphology and dynamics of calcium carbonate granules produced by different earthworm species. *Pedobiologia*, 47, 511–521. Blum, W. (1993). Soil protection concept of the Council of Europe and integrated soil research. In: Integrated Soil and Sediment Research: A Basis for Proper Protection, ed. H.J.P. Eisjackers and T. Hammers, 37-47. *Kluwer Academic Publishers*, Dordrecht.

Chapin, F; Sala, O.E.; Huber-Sanwald, E. (2000). Future Scenarios of Global Biodiversity. *Springer-Verlag*, New York.

Coleman, D.C.; Crossley, D.A.; Hendrix, P.F. (2004). Fundamentals of soil ecology. *Elsevier Academic Press*, Burlington.

Cortet J.; Gillon D.; Joffre R.; Ourcival J-M.; Poinsot-Balanguer N. (2002) Effects of pesticides on organic matter recycling and microarthropods in a maize field: use and discussion of the litterbag methodology. *Eur. J. Soil Biol.*, 38, 261-265.

Curry, J.P. (1994). Grassland invertebrates. Ecology, influence on soil fertility and effects on plant growth. Chapman and Hall.

Déchêne, A.D.; Buddle, C.M. (2009). Effects of experimental forest harvesting on oribatid mite biodiversity. *For. Ecol. Manag.*, 258, 1331-1341.

Dindal, D.L. (1990). Soil biology guide. Wiley.

Ebel, A.; Davitashvili, T. (2007). Air, Water and Soil Quality Modelling for Risk and Impact Assessment. *Springer*, Dordrecht.

Gago-Duport, L.; Briones, M.J.I.; Rodriguez, J.B.; Covelo, B. (2008). Amorphous calcium carbonate biomineralisation in the earthworm's calciferous gland: pathways to the formation of crystalline phases. *J. Struct. Biol.*, 162, 422–435.

Galli, L.; Capurro, M.; Shrubovych, J.; Torti, C. (2012). Phenology of Protura in a northwestern Italian forest soil (Hexapoda: Protura). *Acta Zoologica Cracoviensia*, 55(1), 33-43.

Ghilarov, M.S. (1978). Bodenwirbellose als indikatoren des bodenhaushaltes und von. In bodenbildenden prozesse. Pedologia, 18.

Giordano, A. (1999). Pedologia. UTET.

Hagvar, S. (1982). Collembola in Norwegian coniferous forest soils. 1. Relation to plant communities and soil fertility. *Pedobiologia*, 24, 255-296.

Hansen, B.; Alroe, H.F.; Kristensen, E.S. (2001) Approaches to assess the environmental impact of organic farming with particular regard to Denmark. *Agric. Ecosys. Environ.*, 83, 11-26.

Harmon, M.E.; Franklin, J.F.; Swanson, F.J.; Sollins, P.; Gregory, S.V.; Lattin, J.D.; Anderson, N.H.; Cline, S.P.; Aumen, N.G.; Sedell, J.R.; Lienkaemper, G.W.; Cromack, K.; Cummins, K.W. (1986). Ecology of coarse woody debris in temperate forests. *Adv. Ecol. Res.*, 15, 133-302.

Hassal, M.; Visser, S.; Parkinson, D. (1986). Vertical migration of *Onychiurus subtenius* (Collembola) in relation to rainfall and microbial activity. *Pedobiologia*, 29, 175-182.

Heliovaara, K.; Vaisanen, R. (1984). Effects of modern forestry on northwestern European forest invertebrates-a synthesis. *Acta Forestalia Fennica*, 83, 1-96.

Hoekstra, J.M.; Bell, R.T.; Launer, A.E.; Murphy, D.D. (1995). Soil arthropod abundance in coastal redwood forest: effect of selective timber harvest. *Env. Ent.*, 24, 246-252.

Hole, F.D. (1981). Effects of animals on soil. *Geoderma*, 25, 75-112.

Jabin, M.; Mohr, D.; Kappes, H.; Topp, W. (2004). Influence of deadwood on density of soil macro-arthropod in a managed oak-beech forest. *For. Ecol. Manag.*, 194, 61-69.

Jeffery, S.; Gardi, C.; Jones, A.; Montanarella, L.; Marmo, L.; Miko, L.; Ritz, K.; Peres, G.; Römbke, J.; van der Putten, W.H. editors (2010). European Atlas of Soil Biodiversity. Publications Office of the European Union, Luxembourg.

Killham, K. (1994). Soil Ecology. Cambridge University Press.

Kuznestova, N.A. (2002). Classification of collembolan communities in the east-european taiga. *Pedobiologia*, 46, 373-384.

Lambkin, D.C.; Gwilliama, K.H.; Laytona, C.; Cantib, M.G.; Piearcec, T.G.; Hodson, M.E. (2011). Production and dissolution rates of earthworm-secreted calcium carbonate. *Pedobiologia*, 54S, S119-S129.

Lawton, J.H.; May, R.M. (1995). Extinction Rates. *Oxford University Press*, Oxford.

Lee, M.R.; Hodson, M.E.; Langworthy, G. (2008). Earthworms produce granules of intricately zoned calcite. *Geology*, 36, 943-946.

Leoni, A. (2008). Studio della biodiversità vegetale e del popolamento a microartropodi edafici nella riserva naturale "Guadine Pradaccio". Ph.D. dissertation.

Maharning, A.R.; Mills, A.A.; Adl, S.M. (2008). Soil community changes during secondary succession to naturalized grasslands. *Appl. Soil Ecol.*, 41, 137-147.

Mansourian, S.; Vallauri, D.; Dudley, N. (2005). Forest restoration in landscapes: beyond planting trees. *Springer,* New York.

Marshall, V.G. (2000). Impacts of forest harvesting on biological processes in northern forest soils. *For. Ecol. Manag.*, 133, 43-60.

Menta, C. (2012). Soil fauna diversity – Function, soil degradation, biological indices, soil restoration. In: Biodiversity Conservation and Utilization in a Diverse World. Gbolagade Akeem Lameed (Ed.), InTech, Published.

Menta, C.; Leoni, A.; Gardi, C.; Conti, FD. (2011). Are grassland important habitat for soil microarthropod conservation? *Biodiversity Conservation*, 20, 1073-1087.

Moore, J.D.; Ouimet, R.; Camiré, C.; Houle, D. (2002). Effects of two silvicultural practices on soil fauna abundance in a northern hardwood forest, Québec, Canada. *Canadian Journal of Soil Science*, 82, 105-113.

Parisi, V.; Menta, C.; Gardi, C.; Jacomini, C.; Mozzanica, E. (2005). Microarthropod communities as a tool to assess soil quality and biodiversity: a new approach in Italy. *Agric., Ecos. Env.*, 105, 323-333.

Parthasarathi, K.; Ranganathan, L.S. (1999). Longevity of microbial and enzyme activity and their influence on NPK content in pressmud vermicasts. *Eur. J. Soil Biol.*, 35 (3), 107-113.

Petersen, H.; Luxton, M. (1982). Quantitative ecology of microfungi and animals in soil and litter. *Oikos*, 39, 287-388.

Ponge, J. (1993). Biocenoses of Collembola in atlantic temperate grass-woodland ecosystem. *Pedobiologia*, 37, 223-244.

Ponge, J.F.; Vannier, G.; Arpin, P.; David, J.F. (1986). Caractérisation des humus et des litières par la faune du sol intérêt sylvicole. http://hdl.handle.net/2042/25685 | DOI : 10.4267/2042/25685.

Rusek, J. (2001). Microhabitats of Collembola (Insecta: Entognatha) in beech and spruce forestsand their influence on biodiversità. *Eur. J. Soil Biol.*, 37, 237-244.

Seastedt, T.R.; Crossley, Jr. D.A. (1981). Microarthropod response following cable logging and clear-cutting in the southern Appalachians. *Ecology*, 62, 126-135.

Siddiky, Md.R.K.; Schaller, J.; Caruso, T.; Rilling, M.C. (2012). Arbuscular mycorrhizal fungi and collembola non-additively increase soil aggregation. *Soil Biol. Biochem.*, 47, 93-99.

Sousa, J.P.; da Gama, M.M.; Pinto, C.C.; Keating, A.; Calhộa, F.; Lemos, M.; Castro, C.; Luz, T.; Leitão, P.; Dias, S. (2004). Effects of land-use on Collembola diversity patterns in a Mediterranean landscape. *Pedobiologia*, 48, 609-622.

Taylor, A.R.; Pflug, A.; Schroter, D.; Wolters, V. (2010). Impact of microarthropod biomass on the composition of the soil fauna community and ecosystem processes. *Eur. J. Soil Biol.*, 46, 80-86.

Theenhaus, A.; Schaefer, M. (1995). The effects of clear-cutting and liming on the soil macrofauna of a beech forest. *Forest Ecology and Management*, 77, 35-51.

Wallwork, J.A. (1970). Ecology of soil animals. McGraw-Hill, London.

Wiecek, C.S.; Messenger, A.S. (1972). Calcite contributions by earthworms to forest soils in Northern Illinois. *Soil. Sci. Soc. Am. J.*, 36, 478-480.

Yang, X.; Chen, J. (2009). Plant litter quality influences the contribution of soil fauna to litter decomposition in humid tropical forests, southwestern China. *Soil Biol. Biobhem.*, 41, 910-918.

Zanella, A.; Tomasi, M.; De Siena, C.; Frizzera, L.; Jabiol, B.; Nicolini, G. (2001). Humus forestali. Centro Ecologia Alpina.

In: Woodlands
Editor: Lazaro Manzanares

ISBN: 978-1-62618-177-9

Chapter 3

CLIMATE CHANGE AND FOREST FIRES IN THE AVE REGION (NW OF PORTUGAL)

António Bento-Gonçalves**[*1], **António Vieira**[1], **Flora Ferreira-Leite**[2], **Paula Alexandra Malta**[3] and **Luís da Vinha**[4]

[1]Departamento de Geografia da Universidade do Minho, Centro de Estudos em Geografia e Ordenamento do Território (CEGOT)
[2]Centro de Estudos em Geografia e Ordenamento do Território (CEGOT)
[3]Departamento de Economia, Gestão e Engenharia Industrial da Universidade de Aveiro
[4]Núcleo de Investigação em Geografia e Planeamento da Universidade do Minho (NIGP), Portugal

ABSTRACT

There is a wide array of studies and evidence that climate is changing and these changes will manifest themselves very differently in various areas of the planet.

The weather conditions that occur in Portugal, especially during the summer, are favorable to the occurrence of forest fires. However, the ignition and spread of a fire depends on the interaction of several factors besides the weather, namely the presence of fuel and rugged terrain.

---

[*] E-mail: bento@geografia.uminho.pt.

It is expected that fire regimes immediately respond to climate change, and may even outweigh the direct effects of global warming in the patterns of species distribution and productivity.

The project ADAPTACLIMA - Adaptation to the effects from climate change (INTERREG SUDOE) was based on the preparation of a series of studies on forecasting and analyzing the vulnerabilities and potentialities in Southwest European territories, with the aim of creating a collaborative network of stable institutions. The purpose was to permit both the transmission of knowledge and exchange of experiences among members of the partnership, as well as mutual learning and co-generation of new knowledge. The fundamental task of the network was the preparation of a Plan for Adaptation to Climate Change in Space SUDOEto be implemented in the participating areas. Among the challenges facing SUDOE to mitigate climate change, two of them represent the strategic objectives of the project ADAPTACLIMA: on the one hand, to alert the populations of the SUDOE space to the real consequences of climate change; and, secondly, to promote and develop measures which help adapt society to future scenarios derived from these changes.

In the northwest of Portugal and in particular in the Averegion, one of the main impacts expected from climate change is an increase in number and size of fires and their recurrence.

In terms of plant species, those better adapted to fire will dominate, generating monospecific formations or small variations, wherein all the plant species are of the same age.

Climate change may cause a substantial increase in the risk of fire. In addition, the period of fire occurrence will extend throughout the year, implying a larger fire-fighting organizational structure, which will maintain higher levels of alert for longer periods each year.

INTRODUCTION

Climate Change

Recent decades have been marked by a rising concern with the issues related to climate change and its direct and indirect effects on society. Parallel with this tendency, research seeking to identify causes, effects, and mitigation measureshas also multiplied.

Internationally, this research has been conducted by the IPCC - Intergovernmental Panel on Climate Change, sponsored by the World Meteorological Organization and by the United Nations Program for the Environment.

This organization has been active in understanding the mechanisms that affect climate change, its potential impacts on the planet and on human activities, and in defining strategies of adaptation or mitigation of this phenomenon. It is widely agreed that, based on the data obtained and assessed so far, we have been witnessing a period of global warming and the influence of the anthropogenic factors have helped catalyze this phenomenon.

Analysis of climate data and its trends, as well as the influence of human actions are under scrutiny and modeling, in order to foresee and predict the behavior of the climate system, and to evaluate such climate changes. For this purpose, some physico-mathematical models of climate (Global Climate Models - GCM) have been applied to obtain global trend projections of climate change. These models allow the establishment of different hypothetical socioeconomic scenarios. However, the low spatial resolution of these global models precludes the establishment of detailed regional analyses. Accordingly, methodologies for defining regional scenarios of climate change have been proposed, with a number of examples at the European and Iberian level (Gutiérrez *et al.*, 2010).

ADAPTACLIMA Project

The project Adaptaclima - Adaptation to the effects from climate change - is based on the preparation of a series of studies on forecasting and analyzing the vulnerabilities and potentialities in the Southwestern European territories (Interreg Sudoe), with the aim of creating a collaborative network of stable institutions permitting both the transmission of knowledge and exchange of experiences among members of the partnership, as well as mutual learning and co-generation of new knowledge.

The fundamental task of the network wasto prepare a Plan for Adaptation to Climate Change in Space SUDOE for implementation in the participating areas. Among the challenges facing Southwestern European territories in combating climate change, two of them represent the strategic objectives of the ADAPTACLIMAproject: Firstly, to alert the populations of the SUDOE space to the real consequences of climate change; secondly, to promote and develop measures which will help adapt society to future scenarios derived from these changes.

In the Northwest of Portugal, notably in the Averegion[1] (Figure 1), one of the main impacts expected to result from climate change is an increase in the number and size of fires and their recurrence, and consequently an increase in the erosion of the top soil layer —where the only nutrients available in most Portuguese soils are located (Bento-Gonçalves *et al.*, 2008).

As part of the ADAPTACLIMA project, an evaluation of the magnitude of climate change in various regions of Southwestern Europe, including the Averegion, was conducted. PRUDENCE project outcomes (http://prudence.dmi.dk), which present a series of climate change projections for Europe with a horizontal resolution of approximately 50 km, were initially used.

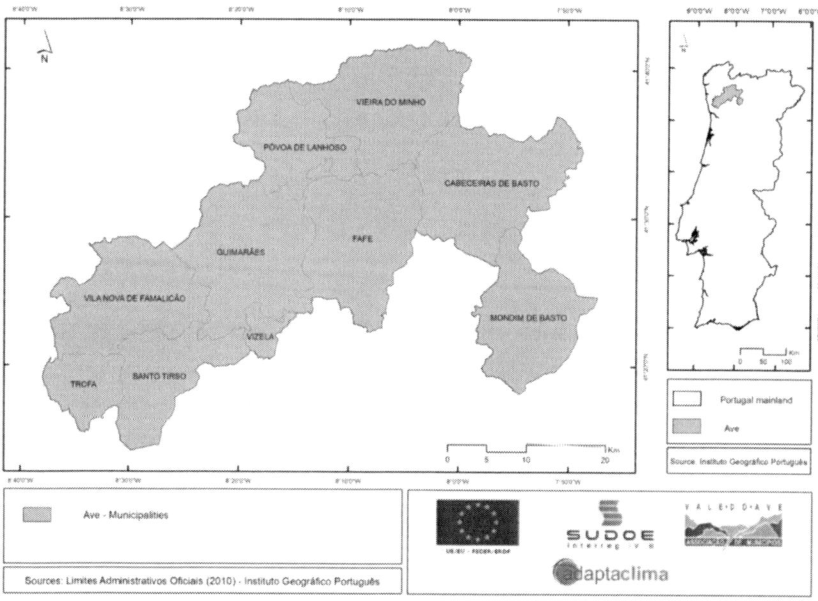

Figure 1. The Ave: administrative framework.

Such projections were made by different institutions utilizing different European regional climate models, based upon the global template HadAM3H,

[1] Located in Minho, more precisely in theNorthwestof Portugal, the Averegionencompasses 10 municipalities (Cabeceiras de Basto - 241,84 Km^2, Fafe - 219,09 Km^2, Guimarães - 241,28 Km^2, Mondim de Basto - 172,09 Km^2, Póvoa de Lanhoso - 132,54 Km^2, Santo Tirso - 136,50 Km^2, Trofa - 71,88 Km^2, Vieira do Minho - 218,48 Km^2, Vila Nova de Famalicão - 201,70 Km^2 e Vizela - 24,70 Km^2), ofthe NUT III Ave andofthe Intermunicipal Community (MIC) ofthe Ave.

which is one of the global models used in the IPCC (Intergovernmental Panel on Climate Change) and the one that offers the best outcomes for the current climate.

A projection for the period of 2071-2100 was established for each regional model, as well as a controle simulation for the period of 1961-1990 which served as basis for the simulation.

After this preliminary analysis, a more detailed investigation for the Averegion was conducted (MeteoGalicia, 2010b), based on the meteorological stations installed in the north-west Portuguese territory.

METHODS AND MATERIALS

Methodology

The development of predictive scenarios of climate evolution for the Ave region (figure 1) was constructed through the analysis of climatic variables based on regionalization techniques that enabled the projection of the IPCC global models to the regional context.

These procedures allow greater accuracy in predicting the specific impacts at the local-regional scale, as well as the implications within the context of climate risks.

In the first stage, an examination of the variability and changes in temperature and precipitation predicted in future climate change scenarios was carried out based on projections of regional climate models. Taking as the starting point the findings of the PRUDENCE project (MeteoGalicia, 2010a), which revealed changes in the projections for the period 2071-2100, two evolution scenarios have been defined: scenarios A2 and B2. These scenarios fall within the typology of evolutionary scenarios defined by the IPCC and take into account the influence of socioeconomic factors on climate change. In a second phase, firmly based upon the scenarios previously defined, a more detailed examination of the Ave region (MeteoGalicia, 2010b), was carried out using data from the meteorological stations in Braga, Pedras Rubras, and Montalegre (Figure 2).

The variables used in the present study correspond to the series of maximum, average, and minimum temperaturesand precipitation.

Using thesemeteorological stations exhibits some problems and raises some methodological issues, which however are not likely to be eliminated.

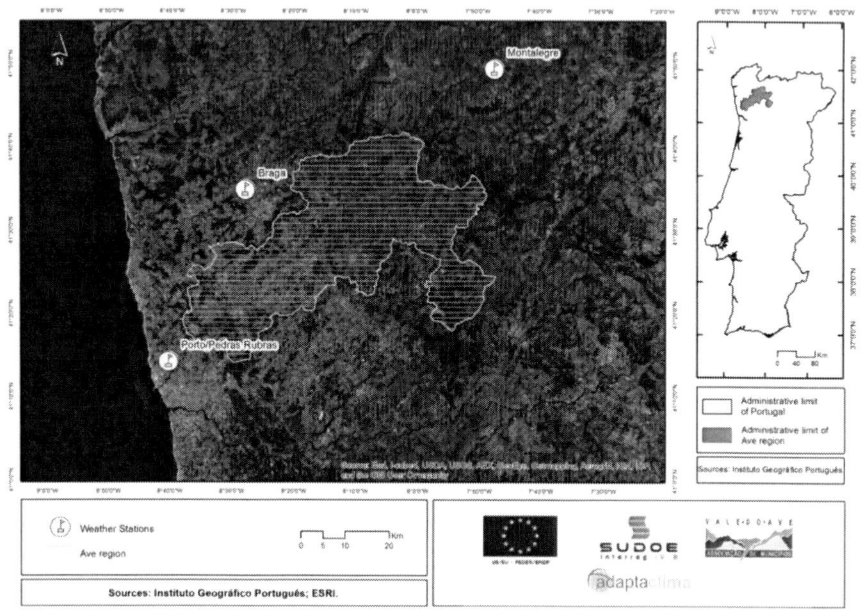

Figure 2. The Ave: geographic location of the meteorological stations employed.

Table 1. The Ave: Meteorological stations used for temperature data

Meteorological stations used for temperature				
List	Station	Initial year	Final year	Station type
1	Braga (Posto Agrário)	1970	2006	Manual
1	Braga (Merelim)	2007	2009	Automatic
2	Montalegre	1970	1999	Manual
2	Montalegre	2000	2009	Automatic
3	Porto/Pedras Rubras	1970	1998	Manual
3	Pedras Rubras	1999	2009	Automatic

Source: MeteoGalicia, 2010b.

The problems are due to the fact that the stations are all located outside the geographical area of the Ave, putting into question the representativity of the data. However, given the low density of the national network of meteorological stations, these stations are the most reliable for expressingthe climatic characteristics of the studyarea.

Another problemencountered depends on the availability of accurate, complete, and consistent data series. Some of these limitations are related to the changesoccurred in the climate registering equipment in all the stations, as well as the change in the location of Braga meteorological station (Table I). In fact, at different times during the study, the measuring devices available at the stations were upgraded and modernized from manual to automatic, with clear implications in terms of the quality of data collected. Moreover, the fact that the upgrades occurred at different times in each station prevents the establishment of correlations between the automatic and manual stations.

Accordingly, there is a clear lack of homogeneity in the maximum temperature series registered in Braga, which is not reproduced in terms of average and minimum temperatures.

In addition, problems related with the quality of station data sets were detected, namely: missing data, particularly noticeable regarding precipitation; presence of monthly accumulated precipitation data, when these should be recorded daily; presence of precipitation records with values under 0 liters. For this reason, an extensive data quality control and homogenization process was conductedby MeteoGalicia, the company responsible for processing the data.

Materials

The Ave Region

The Ave region reveals a significant demographic growth, which has contributed to the establishment of one of the major settlements in Portugal, as well as one of the oldest. Currently, the Ave region has a demographic density of about 406 hab/km^2. Nevertheless, inter-municipal differences show that there is an inverse correlation between the orography and population density. In effect, the population density increases from the eastern part of the region westwards.

The most striking climate feature of the Portuguese Northwest, where the Ave municipalities are located, is unquestionably the high precipitation levels, associated with the frequent passage of frontal surfaces. This feature, coupled with the effects of the mountains located close to the coast, determines an average annual precipitation exceeding 1400 mm.

In fact, the Ave is a region with Mediterranean affinities but with strong Atlantic influence. This results in a climate characterized by mild temperatures, with limited temperature ranges, and heavy average rainfalldue to its geographical positionclose to the Atlantic Ocean. The shape and disposition of the main mountain ranges of the region also contribute to determine this particular climate.

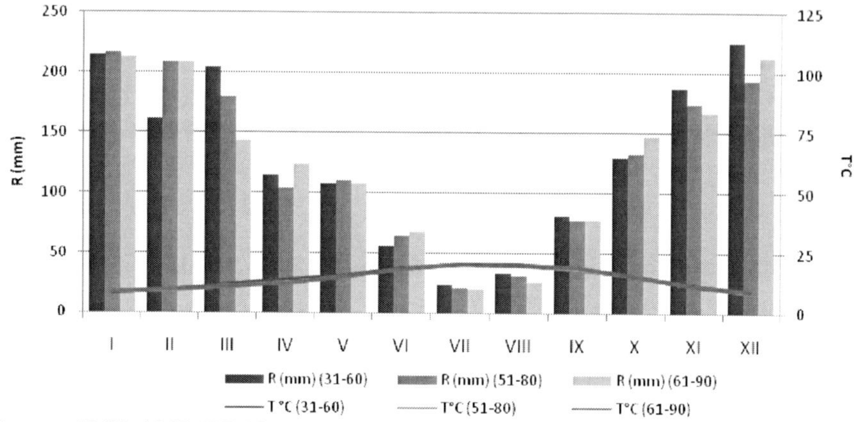

Source: SMN, 1965; INMG, 1991; IM 2005.

Figure 3. Braga's Temperature-Precipitation Graphic (1931-1960 / 1951-80 / 1961-90).

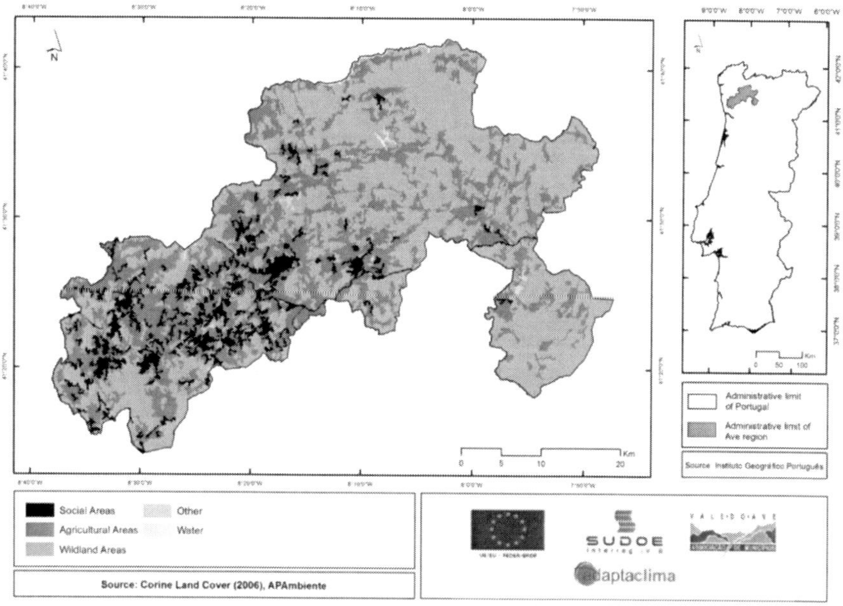

Figure 4. The Ave: land use (2006).

Considering the Climatological Station of Braga located at 41° 33'north latitude, 8° 24' west longitude and at an altitude of 190 meters— it was determined that, in the period 1951-1980, the annual rainfall exceeded 1500 mm (1514.8 mm), distributed over the whole year and counting up 130.4 days with precipitation. The highestaverage monthly temperature was recorded in July (20.2°C), while the lowesttemperature occurs in January (8.7°C). The average annual temperature is around 14°C and the annual temperature range is 12°C (Figure 3).

The Ave region is characterized by the presence of soils with good agricultural capability, high population densities, and high industrial concentration. In terms of land use, the region displays a widespread distribution of the different uses throughout the entire landscape (Figure 4).

The forest area in the Ave region is comprisedof 45.6% of maritime pine, 35.1% eucalyptus, and 9.1% oaks (Figure 5). Due to the fact that the region is heavily humanized, the organization of the landscape presents sharp contrasts. This does not however preclude the existence of an ecologicallyunitary system.

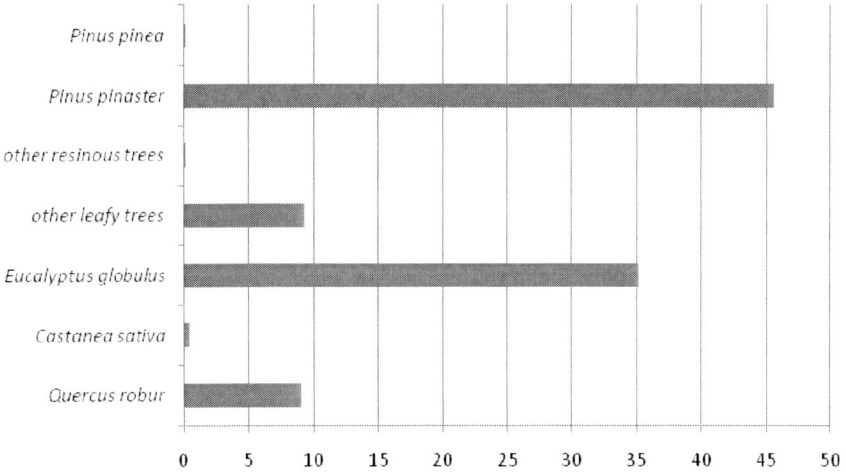

Source: AFN.

Figure 5. The Ave: distribution (%) of the forest by forest species.

In fact, the region distinguishes itself nationally due to the unique uniformity of its landscape physiognomy, of the set of its inhabitant species, and by its particular adaptive strategies. Thus, the geographical area corresponding to the Ave presents sharply contrasting landscape systems, resulting from a long and intense human occupation, constrained by particular physical conditions, both in terms of topography and climate.

Being a fundamental element of the landscape, the vegetation is an outstanding testament to the soil and climate conditions, as well asto the anthropic action in a particular territory.

In this geographical context, six basic types of land use can be distinguished based on the basic landscape units (GOMES 2001):

- plain farming systems, corresponding to the lowland landscape of scattered settlements;
- mountain farming systems, where villages are grouped;
- urbanized systems;
- uncultivated systems, mostly composed of more or less degraded scrublands and regions of skeletal soil and bare rock;
- plantation forests with predominance of maritime pine *(Pinus pinaster)* and eucalyptus *(Eucalyptus globulos)*;
- forests of deciduous species, dominated by two types of oaks - *Quercus robur* and *Quercuspyrenaica*, corresponding to spontaneous or sub-spontaneous areas.

Forest Fires in the Ave Region

The evolution of the number of forest fire occurrences and burnt areas in the Ave (considering all 10 municipalities studied), between 1980 and 2009, reveals alarge number of deflagrations translated into a total of 56473 occurrences and a total of 108836 hectares of burnt area.

The annual change in the number of occurrences within the period considered (1980-2009) reveals that while the 1980s recorded the lowest values, from the end of the decade, more precisely in 1989, there wasa rise of about 93% in comparison with the previous year. From thatyear on, the number of occurrences was always above 1000.

It should be noted that there is a positive correlation between the number of occurrences and the amount of burnt area throughout the time period in analysis (Figure 6), although this correlation is higher in terms of occurrences (R^2=0,5).

In terms of the burnt areas in the Ave region (1980-2009), the 1980s also display the lowest values recorded. In effect, the years 1983, 1986, and 1988 register the lowest values of the series (157, 32, and 354 hectares).

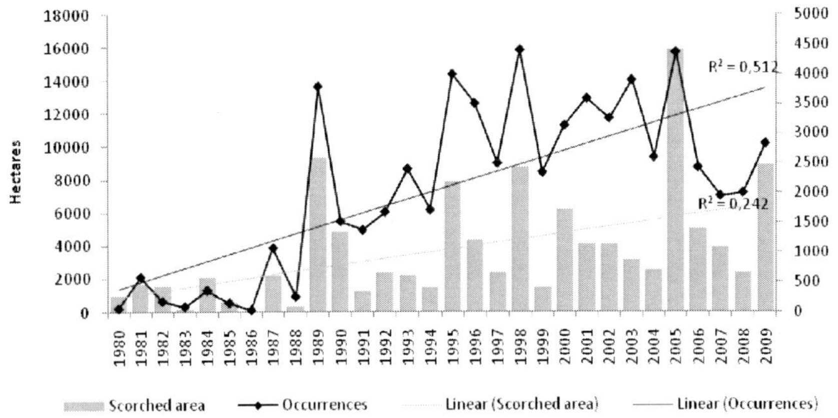

Source:AFN.

Figure 6. The Ave: evolution of the number of forest fire occurrences and burnt areas (1980-2009).

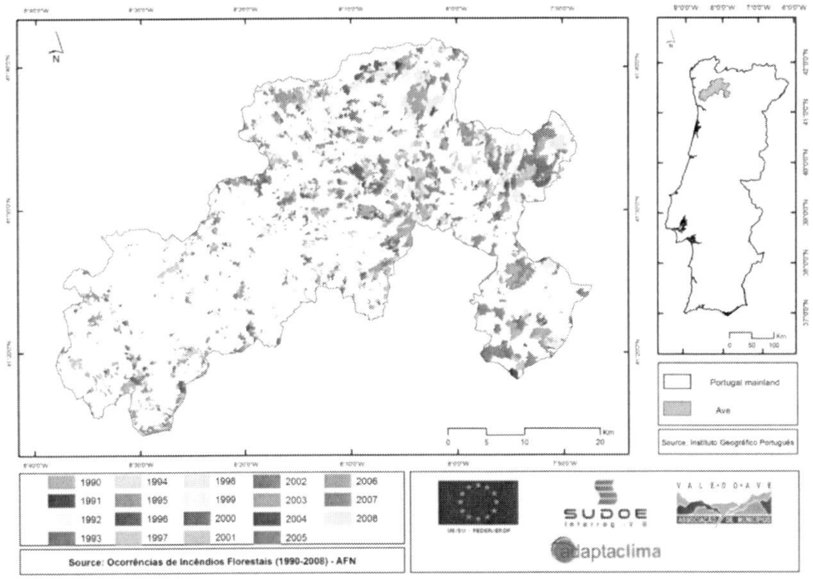

Figure 7. The Ave: spatial distribution of the forest fires occurrences (1990-2008).

The year 1989 stands out as the exception (9150 hectares) and marks the turning point in the evolution of the burnt areas. It is worth noting the declining trend in the burnt areasfor the years 2006, 2007, and 2008, each with values under 5000 hectares. The year 2009, with a total of 8854 hectares of burnt area, however, counters this trend (Figure 6).

Analysing the data on the burnt areas it is possible to verify an uneven spatial distribution. The most striking feature is the stark contrast between the coastal and inland municipalities, which becomes clearly visible when analyzing the cartography of the burnt areas by municipality for the period between 1990 and 2008.

The municipalities of the interior - hilly, demographically fragile and characterized by a predominance of wild spaces over the remaining land uses - have more extensive burnt areas, as are the cases of the municipalities of Mondim de Basto, Cabeceiras de Basto, Vieira do Minho, Póvoa Lanhoso, and Fafe (Figure 7).

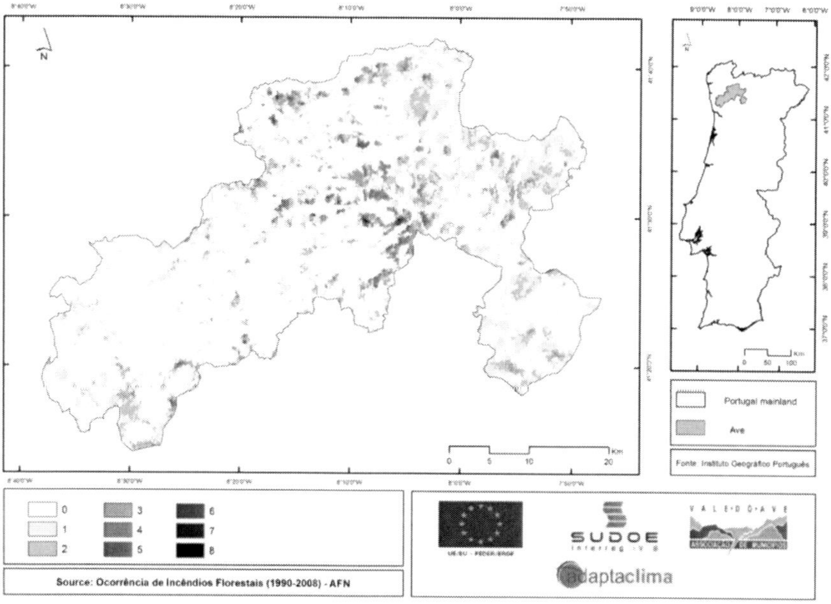

Figure 8. The Ave: recurrence of forest fires.

It is also worth noting that in the 19year period under review, the Ave region displays areas that were repeatedly affected by forest fires. Some areas in the region registered fires on 8 different occasions during thisperiod (Figure 8). In terms of recurrence of forest fires, this is the maximum value observed nationally when analyzed at a homologous scale and with the same cartographic information.

RESULTS

The analysis based on the PRUDENCE project shows that scenario A2 foresees a major increase in the number of emissions due to the rapid and continued growth of global population and to an economic growth at regional level, while scenario B2 presupposes a smaller increase in emissions associated with moderate population growth and slower levels of economic and technological development.

The defining process of the AVE scenarios employed series of temperature (maximum, average, and minimum) and precipitation projected for the period 2071-2100. From this data, temperature and precipitation trends were established, considering their annual, seasonal, and monthly occurrence, as well as the extreme values.

Therefore, scenario A2 points out that the temperature reveals a positive trend, with a forecast of growth between 1.5°C and 5°C in terms of annual average temperature. These figures indicate an average temperature rise of about 0.56°C per decade for the Averegion in the period 2071-2100. In seasonal terms, positive anomalies between 1.5°C and 8°C are observed, especially in the summer months.

As for precipitation, there is an overall decrease of its occurrence. Seasonally, this reduction is more sensitive in the summer, while registering an increase during the winter months. In accordance with these trends, there is a decrease in the frequency of rainy days while the number of days of very severe rainfall rises.

Temperature predictions indicate an augment in the number of hot days and nights, translated in scenario A2 in an increase of 100 days per year. Similarly, a slight decrease in the number of days and cold nights is foreseen. For scenario B2, predictions also point to a rise in the average annual temperature between 1.5°C and 4°C, following however a less pronounced trend. The seasonal analysis of temperatures also demonstrates positive anomalies, higher in the summer, but less significant than in scenario A2.

Precipitation forecasts for scenario B2 are similar to those of scenario A2, registering a general annual decline. Seasonally, there is a more significant reduction during the summer, while the winter months record an increase.

Another important dimension is the decrease in the frequency of rainy days and the extension of very intense rainy days, though with levels below scenario A2 for both variables. For extreme temperatures, scenario A2 predicts an increase in the number of hot days and nights (more 86 days per year) and a slight decline in the number of cold days and nights, though less pronounced.

Following such analytical procedures, a more detailed study focusing the Ave region was developed. The findings derived from the implementation of regional climate models allow for some conclusions regarding the temperature and precipitation trends.

Temperature

Through the analysis of the temperature series, the annual, seasonal, and monthly trends for the period of 1970-2009 were identified.

Examining the annual trend, a significant increase in temperature is recorded. For the Braga and Montalegre stations, the increase in the maximum, mean, and minimum temperatures is 0.5°C per decade.

Data on the Pedras Rubras station indicates a rise of 0.5°C per decade for the mean and minimum temperatures, but of just 0.2°C perdecade for the maximum temperature.

The behavior of the temperature trend is similar in the various stations, whereassuch behavior is not homogeneousif the time period filter is applied. In fact, assessing this trend through time, i.e. throughout the reference period for this survey, reveals that the sharp temperature increase dates back just to the mid 1970s.

Regarding the seasonal trend, spring is the season of the year which presents the largest increase in temperature, reaching values around 0.7°C per decade. This behavior occurs in all the series and temperatures considered, albeit with less intensity.

The monthly temperature trend highlights March as the month with the most significant increase: around 1°C per decade is the increase recorded in the Braga and Montalegre stations, while in Pedras Rubras the numbers are less significant, never exceeding 0.6°C per decade.

Based on daily data of the three temperature series, another relevant aspect concerns the analysis of extremes[2]. This particular analysis shows a significant decline in the frequency of cold days and cold nights. This trend is more pronounced in the spring and summer and reaches the most significant values

[2] For the analysis of temperature extremes, the methodology applied included the calculation of the percentiles p95 and p5 for the maximum and minimum temperatures in each series and for each season. Hot days designate those in which the maximum temperature exceeded the 95 percentile and hot nigths those in which the minimum temperature exceeded the 95 percentile. Similarly, cold days are those in which the maximum temperature was below the 5 percentile and cold nights the ones registering a minimum temperature below the 5 percentile.

in Pedras Rubras during the summer months (a decrease in cold nights of 3.34 nights/decade).

As for the number of hot days, the increase mainly occurs in the spring and summer seasons. This rise is around 1.5 days/decade during the spring (more significant in Braga) and about 2.5 days/decade in the summer (Montalegre and Braga). A rise in the number of hot nights was also identified. In the case of Braga, this increment takes place not only in the spring and summer, but also in the autumn, while in Pedras Rubras it occursall year long.

Precipitation

A homologous procedure to the one performed for temperature allowed for the identification of annual, seasonal, and monthly precipitation trends for the period 1970-2009, for the Braga, Montalegre, and Pedras Rubras stations (Table II).

Concerning the accumulated annual precipitation, no significant trend in any of the three series is observable. As for the seasonal behavior of precipitation, a nearly significant positive trend, corresponding to a likely increase of 1.58% per decade, is recorded during the autumn season in Braga. The same pattern occurs in Montalegre where a nearly significant positive trend associated to a likely increase of 1.97% per decade emerges in the autumn seasons.

Shifting to a monthly scale, a precipitation decrease in February and an increase in October are forecasted. These results are consistent with those obtained in the study conducted by Cruz *et al.* (2006) on the trends for Galicia, for the period of 1960-2006. Finally, regarding the rainfall extremes, no detectable annual trend is drawn from the examination of the frequency of rainy days in which certain limits are overcome.

Table 2. The Ave: meteorological stations used for precipitation

Meteorological stations used for precipitation				
List	Station	Initial year	Final year	Station type
1	Braga (Posto Agrário)	1970	2006	Manual
1	Braga (Merelim)	2007	2009	Automatic
2	Montalegre	1970	2009	Manual
3	Porto/Pedras Rubras	1970	2009	Manual

Source: MeteoGalicia, 2010b.

Discussion

Once the major trends in terms of climate change for the Ave region were surveyed, the following step consisted in identifying the primary and secondary impacts to the forest (Table III).

As a consequence of the rise of 0.2 to 0.5°C per decade of the maximum, average, and minimum temperatures, an increasein forest fire risk is estimated, as well as an extension of the "fire season". This new scenario will be characterized by a larger number of occurrences, higher recurrence of fires and the occurrence of more and larger "big forest fires".

Conditions for the increase of burnt areas are created with implications that will result in a devaluation of the landscape and a loss of its touristicattraction, as well as increased erosion, changes in land use, loss of biodiversity (spread of invasive non-native species), and decreased productivity. The expected increase of the average temperature of 0.6°C to 1°C in the month of March, combined with the decrease of precipitation in February, will be reflected in a rise of the forest fire risk during the month of March. Accordingly, a significant increase of the recurrence of forest fires during this month is expected. Consequently, an enlargement of the burnt areas, erosion, and forest-grazing conflicts will take place. The report prepared by MeteoGalicia points to an increase of hot days in the spring (1.5 days/decade) and during the summer months (2.5 days/decade), as well as an increase of hot nights in the autumn, which will catalyse conditions favorable to an increase in the number of heatwaves.

This phenomenon will ultimately entail an increased risk of forest fires, increasing number of occurrences, higher recurrence of fires, and the occurrence of more and larger "big forest fires". Within this predictive framework, theautumn will present favorable conditions for the occurrence of "big forest fires".

These conditions will generate a rise of the burnt areas, which can imply a devaluation of the landscape and the loss of its touristic attraction, increased erosion (with the impoverishment of the soil), and change in land usewith loss of biodiversity (spread of invasive non-native species), and decreased productivity.

Finally, a predictable increase in precipitation in the autumn, especially during the month of October, should be noted. This is theseason when soils are often unprotected of vegetation as a result of forest fires. This situation can lead to an impoverishment of soil nutrients, involving changes in land use, loss of biodiversity, and decreased productivity.

Table 3. Climate Change and its impacts on the Ave forest

Climate Changes	Primary Impacts	Secondary Impacts
Temperatures Maximum, average and minimum temp. rise of 0.2 to 0.5°C/decade	- increased risk of forest fires - forest fire season extends over more months - more forest fires - higher recurrence of forest fires - more "big forest fires"	- increase of burnt areas - devaluation of the landscape - loss of tourism attraction - increased erosion - change in land use - loss of biodiversity - decrease in productivity
Average temperature significant increase in the spring March – rises 0.6 to 1°C Precipitation decline in February	- more forest fires in March - higher recurrence of forest fires	- increase of the burnt areas - increased erosion - increased forest-grazing conflicts
Hot days significant increase in Spring and Summer 1.5 days/decade in Spring 2.5 days/decade in Summer	- increase of heatwaves - increased risk of forest fires - more forest fires - higher recurrence of forest fires - more "big forest fires"	- increase of the burnt areas - devaluation of the landscape - loss of tourism attraction - increased erosion - change in land use - loss of biodiversity - decrease in productivity
Hot nights significant increase Braga – in the Autumn Pedras Rubras - in all seasons	- increase of heatwaves - increased risk of forest fires - more forest fires - higher recurrence of forest fires - more "big forest fires" - large forest fires extend over the autumn	- increase of burnt areas - devaluation of the landscape - loss of tourism attraction - increased erosion - change in land use - loss of biodiversity - decreasein productivity
Precipitation nearly significant increase in Autumn Braga – 1.58%/decade Montalegre - 1.97%/decade increase in Octobre	- impoverishment of soils	- change in land use - lossof biodiversity -decrease in productivity

The scenario described above grounded the preparation of a Forest Adaptation Plan to Climate Change in the Ave. This plan presents 6 major measures which are compatible with the measures advocated in the Municipal Plans for Forest Fire Protection, which aimed at contributing to the major goals set in the National Plan for Forest Fire Protection (Lourenço, 2005):

I. Elaboration of managing plans of mountain areas

Goals: - Increasing the territorialresilience to forest fires.
- Reducing the incidence of forest fires.

The use of fire in mountain areas should be subject to a tight control by regulating cattle herding and the rationalization of agricultural burning, which may be replaced by controlled fires. This may involve the preparation of intermunicipal spatial plans, targeting the territorial management of mountain ranges and the regulation of land uses. Such plans must take into account the Forest Management Plans, the Plans of Wasteland Use and the Municipal Plans for Forest Fire Protection. Additionally the plans should seek the involvement of all local and regional actors, i.e., actively seeking to involve the communities themselves through their empowerment, by reducing mistrust/disputes, achieving broad consensus, and establishing land use, livestock, hunting, and burning regulations.

II. Production of legislative measures for the areas of urban-forest interface.

Goals: - Increasing the territorial resilience to forest fires.
- Reducing the incidence of forest fires.

The National Plan for Forest Fire Protection (PNDFCI) defines, as one of its operational objectives, the protection of theurban-forest interface zone in order to increase the resilience of the territories to forest fires.

The Ave region presents a population density exceeding 400 hab/km^2 and a scattered urban territorial model characterized by diffuse patterns of urbanization and industrialization where the multi-functionality of land use interconnects. However, despite the existence of rules and regulations for the protection of forests against fires in close proximity to homes and urban areas, there are legal lacunae that prevent more efficient interventions, especially in terms of prevention and inspection.

III. Implementation of training initiatives for those involved in the prevention, surveillance, and combat of forest fires.

Goals:
- Reducing the incidence of forest fires.
- Improving the efficiency in the combat and management of fires.
- Adaptation of an effective organizational and operational structure.

In a probable scenario of more and larger big fires and of fires with higher intensity and greater severity, issues such as prevention, surveillance, and early detection are critical, just as rapid first intervention is vital. It is therefore crucial to prepare the various agents to this new reality, increasing the effectiveness of prevention and the efficiency of surveillance and combat activities, which should integrate a specific training program for the Ave.

IV. Promotion of technical-scientific studies on emergency measures for soil protection following forest fires

Goal:
- Recovery and rehabilitation of ecosystems.

A set of emergency techniques applied globally to minimize the effects of erosion immediately after the occurrence of forest fires are currently accessible. However, coupled with the scarcity of this type of studies in Portugal, there is no information available for the Ave region on the effectiveness, the cost-benefit relation, or the positive and negative impacts of the implementation of those techniques.

Against such a background of longer, more extensive and severe forest fires, two critical issues arise: not only those of prevention and combat but also aspects of the mitigation of its consequences. In this context, high-priority measures should consider conducting technical and scientific research about the measures to protect soils after forest fires, as well as the promotion of partnerships between universities and public and private regional institutions.

V. Creation of a network of forest meteorological stations, allowing for the implementation of a warning system

Goals:
- Reducing the incidence of forest fires.

- Improving the efficiency in the combat and management of fires.

One of the weaknesses detected in the prevention and combat of forest fires is the absence of meteorological stations within the Ave territory, which precludes the existence of regionally disaggregated weather information in real time.

The creation of a monitoring system of climate conditions to prevent the occurrence of forest fires is thus proposed. This system should include, as a tool for the prevention and combat of forest fires, the automatic collection of climatological information, its treatment, and online access. Moreover, such a system should be constructed upon a network of meteorological stations preferably installed in forest environments and properly located to cover the entire territory of the Ave region.

VI. Development of environmental awareness initiatives for the population, with particular emphasis on the school children

Goals: - Reducing the incidence of forest fires.

Increased risk of forest fire implies a change in the behavior and attitudes of the population.

Therefore, awareness campaigns targeting several different publics, particularly school children, should be promoted in order to obtain lasting outcomes in the medium and long run. Guided by educational, environmental, and spatial (forests) objectives, such campaigns should aim to contribute to the development of the civic education of young people, instilling concepts, principles, values, and attitudes that enable a sustainable use of the forest environment.

Wholesome conviviality and the awareness and education of youth can contribute to a change of attitudes towards forest and its importance, a goal that everyone must commit to in order to preserve, protect, and conserve forest areas.

CONCLUSION

The main goal of ADAPTACLIMA is focused in drafting anadaptationplan to the impact of climate change on forest fires in the Ave region. In this sense, it became imperative to undertakean analysis of the multiple climatic variables and to perform an assessment of the magnitude of climate change in a specific future time frame for the Ave region. Accordingly, climate models based upon regional climate variables were implementedand data analysis was performed in partnership with Meteo Galicia (MeteoGalicia, 2010a, 2010b). Results obtained are consistent with those presented by several authors, both at the national (Santos et al., 2002) and Iberian level (Gutierrez et al., 2010) and which indicate a rise in the annual temperature trend and an increase in hot days and nights. Regarding precipitation, significant trends can only be detectable at the seasonal and monthly level, highlighting a decline in February and an increase in October.

In conclusion, the Ave appears to be territory highly prone to forest fires. This fact, coupled with the expected climate change scenarios might lead to a significant increase in the risk of forest fires and the extension of the fire season, as well as an increase in the size, intensity, and severity of forest fires. Having identified the major trends in terms of climate change, as well as the primary and secondary impacts on the forest spaces of the Ave, an a adaptation plan to climate change was set up. The goals of the plan are articulated in order to enhance the resilience of the territory to forest fires, reverse fire incidence, improve the efficiency of combat and management of fires, restore and rehabilitate ecosystems, and adjust an effective organizational and functional structure.

ACKNOWLEDGMENTS

The authors would like to thank AMAVE (Associação de Munícipios do Vale do Ave), José Martins, Domingos Silva and Vera Soares.

REFERENCES

Bento-Gonçalves, A., Costa, F. (2002) – O Vale do Ave – sua Geografia. In José Mendes, M. Amado, Fernandes, Isabel Maria (Coord. Centíficos) - Património Industrial no Vale do Ave – Um passado com futuro. *Rota do Património do Vale do Ave e ADRAVE*, V. N. Famalicão, pp.40-58.

Bento Gonçalves, A., Vieira, A., Ferreira, A., Coelho, C. (2008) — Caracterização geomorfológica e implementação de um sistema integrado de informação, em ambiente SIG, no âmbito do projecto RECOVER (Estratégias de remediação de solos imediatamente após incêndios florestais). *Revista Geografia Ensino ePesquisa*, V. 12, nº 1, Santa Maria, Rio Grande do Sul, Brasil, pp.3721-3735.

Bento-Gonçalves, A., Vieira, A., Ferreira-Leite, F., Martins, José, Silva, Domingos, Soares, Vera (2011a) – ADAPTACLIMA - Adaptation to the effects from climate change in the AVE. In: Proceedings of the 3rd international meeting of fire effects on soil properties (FESP III), Edited by António Bento-Gonçalves, António Vieira. Guimarães, NIGP-Universidade do Minho, CEGOT, pp. 175-180. *http:// repositorium. sdum.uminho.pt/bitstream/1822/12300/1/FESP_2011_Proceedings.pdf.*

Bento-Gonçalves, A., Vieira, A., Ferreira-Leite, F., Martins, J., Silva, D., Soares, V. (2011b) – Adaptaclima: Adaptação aos efeitos derivados das alterações climáticas. As Mudanças Climáticas e os Incêndios Florestais no Ave. AMAVE, InterregSudoe IV B, ISBN: 978-989-95470-4-9; Guimarães, 94p.

Bento-Gonçalves, A., Vieira, A., Úbeda, X. and Martin, D. (2012) – Fire and soils: key concepts and recent advances. Geoderma, 191, Elsevier, pp. 3-13. (doi: dx.doi.org/10.1016/j.geoderma.2012.01.004).

Cruz, R.; Lago, A.; Lage, A.; Rial, M. E.; Diaz-Fierros, F.; Salsón, S. (2009) – Evolución recente do clima de Galicia. Tendencias observadas nas variablesmeteorolóxicas. In *Evidencias e impactos do cambio climático enGalicia.* Xunta de Galicia, pp. 1-58.

Ferreira-Leite, F., Bento-Gonçalves, A., Vieira, A. (2011) - Therecurrenceintervalofforestfires in Cabeço da Vaca (Cabreira Mountain - Northwestof Portugal). *Environmental Research,* 111, 2, Elsevier, pp. 215-221.

Gutiérrez, J. M., Herrera, S., San-Martin, D., Sordo, C., Rodriguez, J. J., Frochoso, M., Ancell, R., Fernandéz, J., Cofiño, A. S., Pons, M. R., Rodrigues, M. A. (2010) -Escenarios regionales probabilísticos de Cambio Climático en Cantabria: termopluviometría. Gobierno de Cantabria, Consejería de Medio Ambiente. Santander, 108 p. + 60 mapas.

Lourenço, L. (Coord.) (2005) – Plano Nacional de Defesa da Floresta contra Incêndios (PNDFCI). Agência para a Prevenção de Incêndios Florestais (APIF), Miranda do Corvo, 2 vol, 236 p. e anexos.

MeteoGalicia (2010a) – Informe sobre impactos sobre el Val do Ave – Portugal – Variables: precipitación, temperatura, evaporación, viento y radiación de onda corta (relatório interno). Conselleria de Medio Ambiente Territorio e Infraestruturas, Santiago de Compostela,24 p.

MeteoGalicia (2010b) – Informe sobre evidencias en Portugal – Variables: temperatura e precipitación (relatório interno). Conselleria de Medio Ambiente Territorio e Infraestruturas, Santiago de Compostela,56 p.

Santos, F., Forbes, K.,Moita, R. (ed.) (2002) - Climate change in Portugal. Scenarios, ImpactsandAdpatationMeasures – SIAM Project, Gradiva, Lisboa, 454 p.

Serviço Meteorológico Nacional (1965) – O clima de Portugal. Normais climatológicas do Continente, Açores e Madeira, correspondentes a 1931-1960. Fascículo XIII, 207p., Lisboa.

Tavsanoglu, Ç.,Úbeda, X. (2011) - Fire and soils: Methodological issues and implications to management. *Environmental Research* 111, Elsevier, pp. 191-192.

Vieira, A., Bento-Gonçalves, A., Lourenço, L., Martins, C., Ferreira-Leite, F. (2009) - Risco de incêndio florestal em áreas de interface urbano-rural: o exemplo do Ave. Territorium 16,pp. 139-146. *http://www1.ci.uc.pt/nicif/riscos/downloads/t16/interface_urbabo_rural.pdf*

Sources

Atlas do Ambiente (1931-1960), Instituto do Ambiente. Disponível em <*http://www.iambiente.pt/atlas/est/index.jsp*>.

Autoridade Florestal Nacional – Estatísticas de Fogos Florestais. Disponível em <*www.afn.min-agricultura.pt*>.

Autoridade Florestal Nacional – Cartografia de Fogos Florestais. Disponível em <*www.afn.min-agricultura.pt*>.

Autoridade Florestal Nacional – 5º Inventário Florestal Nacional. Disponível em <*www.afn.min-agricultura.pt/portal/ifn/relatorio-final-ifn5-florestat-1*>.

Instituto de Meteorologia (2005) – Normais Climatológicas, Braga – Posto Agrário (1961-1990), Lisboa.

Instituto do Ambiente (2006) – Programa Nacional para as Alterações Climáticas, Anexo Técnico FLORESTA.

Instituto Geográfico Português, Disponivel em <*http://www.igeo.pt*>

Instituto Nacional de Estatística – Recenseamento Geral da População (1991, 2001). Disponível em <*http://www.ine.pt*>.

Instituto Nacional de Estatística – Estimativas de População Residente (2008). Disponível em <*http://www.ine.pt*>.

Instituto Nacional de Meteorologia e Geofísica (1991) – Normais Climatológicas da Região Entre Douro e Minho e Beira Litoral (1951-1980) – O Clima de Portugal, Fascículo XLIV, Volume I, 1ª Região, Lisboa.

In: Woodlands
Editor: Lazaro Manzanares

ISBN: 978-1-62618-177-9

Chapter 4

DRIVERS OF DEFORESTATION AND POTENTIAL FOR CARBON TRADING IN MIOMBO WOODLANDS

***Royd Vinya*[*1], *Stephen Syampungani*[1], *Exhildah C. Kasumu*[1], *Robby Kasubika*[1] and *Concilia Monde*[1,2]**

[1]School of Natural Resources, Copperbelt University, Kitwe, Zambia
[2]Department of Aquatic Ecology and Water Resource Management, Wageningen University, The Netherlands

ABSTRACT

Africa's Miombo woodlands are vast, of high conservation value and essential for human well-being. This forest type plays a pivotal role in livelihoods of the majority of resource-poor rural communities in southern Africa. Majority of these communities have limited access to sources of employment and domestic income because they have little cash or other resources. It is, however, unfortunate that livelihood factors such as agricultural production and use of fire have constantly altered miombo woodlands. These livelihood factors have all profoundly contributed to degradation and in some cases depletion of the forest resources in southern Africa. Many studies have demonstrated that

[*] Corresponding author email: royd.vinya@gmail.com.

deforestation in miombo woodlands is one of the major contributors to global climate change. However, the question of what exactly drives woodlands deforestation at the regional scale remains largely unanswered. Over the last three decades, significant amount of evidence has coalesced around the general assumption that factors that drive tropical deforestation interact in a complex way and fluctuate significantly at global, regional and country levels. Not surprising all UNFCCC negotiations and key debates have tended to revolve around forests. This is mainly because of the realization that the objectives of mitigating climate change may not be reached without extensive action on deforestation and forest degradation. Therefore, a review of drivers of deforestation in miombo woodlands will help in deepening our understanding of what drives miombo woodlands deforestation and what the potential for future carbon trade is like.

1. Introduction

Miombo woodland is the most extensively distributed seasonally dry tropical forest, covering much of southern Africa (Figure 1). The woodland is vast, of high conservation value and supports millions of rural livelihoods in the SADC region.

This forest type plays a pivotal role in the livelihoods of the majority of resource-poor rural communities who have limited access to sources of subsistence income and employment opportunities. It is, however, sad to note that livelihood factors such as wood extraction, agricultural production, infrastructure development and use of fire have all constantly altered this prestigious tropical biome (Chidumayo, 1997). These anthropogenic factors have all profoundly contributed to degradation, and in most cases complete loss of forest cover in the Miombo woodlands (FAO 2005; Lusambo et al. 2007; Syampungani et al. 2009).

Although it has been demonstrated by many authors that tropical deforestation is one of the key drivers of global climate change (Fearnside and Laurance 2003; Henry et. al 2010), factors that drive tropical deforestation at both regional and global scales remain scarcely understood. It is, however, fascinating to note that in the last few decades, evidence has coalesced in support of the notion that, factors that promote tropical deforestation are complex, and varies in scale at global, regional and country levels.

Given the pivotal role that forests play in mitigating climate change, it is not surprising that most of negotiations in UNFCCC are centered on tropical forests.

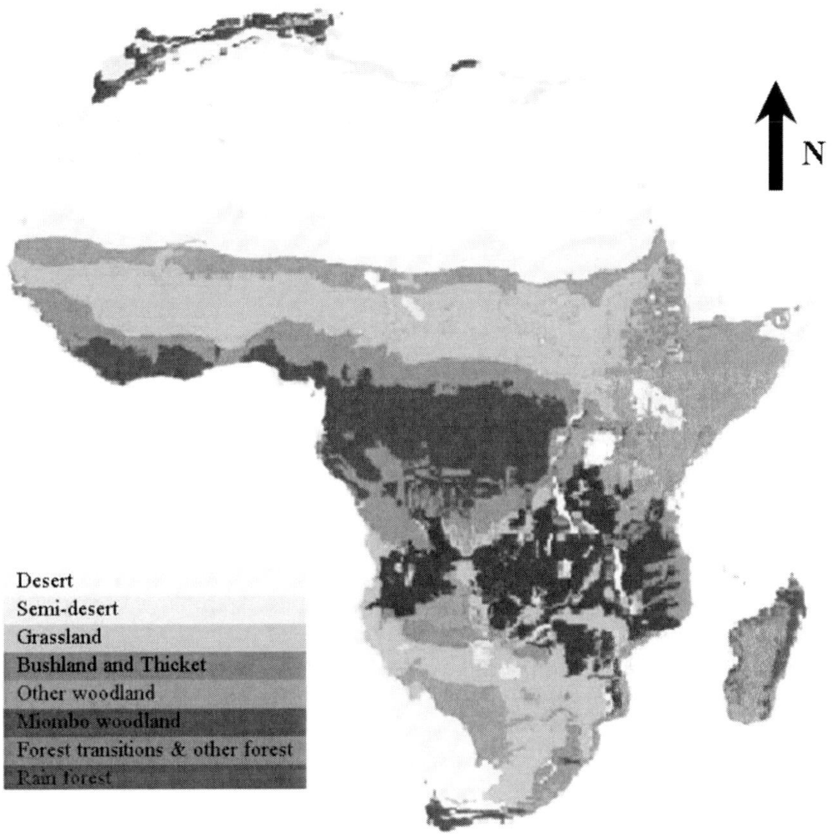

Source: Ryan (2009).

Figure 1. White (1983) original distribution of African ecosystems.

Ideally the key objectives of mitigating climate change may not be attained in the absence of extensive and consented effort in combating tropical deforestation and forest degradation.

2. Deforestation Hotspots and Trends in Miombo Woodlands

Deforestation hotspots in miombo eco-region have been shifting with time. The period between 1990 and 1995 saw Zambia being the centre of significant forest cover loss (Table 1).

Table 1. Mean annual forest cover change for the period 1990 to 2005 in the Miombo eco-region

	Forest area (1,000 ha)		Annual Forest change (%)			
	1990	2005	1990 – 1995	1990 – 2000	2000-2005	1990 – 2005
Angola	23,385	59104	-1	-0.2	-0.2	4
Malawi	3,612	3402	-1.6	-0.9	-0.9	-0.4
Mozambique	17443	19262	-0.7	-0.3	-0.3	0.6
United Republic of Tanzania	34123	35257	-1	-1	-1.1	0.2
Zambia	32720	42452	-8	-0.9	-1	1.5
Zimbabwe	8960	17540	-0.6	-1.5	-1.7	3.3
Regional Average			-2.2	-0.8	0.9	1.5

Source: FAO 1995; 2009.

Mean annual forest cover change for Zambia was 8% which was the highest in the region. On the contrary, Zimbabwe and Mozambique had the least annual rate of forest cover loss. Surprising, Zimbabwe became a deforestation hotspot for the period 1990 to 2000 as it experienced 1.5% annual forest cover loss. A much long term review for the period 1990 to 2005 shows that, while the rest of the countries in the miombo eco-region experienced a positive forest cover change, Malawi remained to be the main deforestation hotspot.

The annual rates of deforestation in miombo eco-region varied from country to country (Table 1). Across the region, the rate of forest cover change was higher between the period 1990 and 2005. Over a ten year period, from 1990 to the year 2000, the region experienced a substantial reduction in annual forest cover loss.

The same can also be said for the years 2000 and 2005. It is interesting to note that over wider span, from 1990 to 2005, there is evidence to suggest that woodlands experienced a significant forest recovery.

The period 2000 and 2005 coincides with increased shift in policy direction in the miombo eco-region. This period represents years when community participation in forest management had gained some momentum throughout the region.

3. What Drives Miombo Woodlands Deforestation?

The proximate drivers of deforestation in miombo woodlands can broadly be categorized into institutional, socio-economic and economical factors. These factors interact in a diverse way to influence forest cover change.

3.1. Institutional Based Factors

3.1.1. Legislative and Policy Environment

Countries in Miombo eco-region have introduced a number of policies and pieces of legislation, aimed at addressing one or more aspects of natural resource management (NRM). Despite their positive contribution to sustainable NRM, these policies demonstrate widespread implementation and enforcement failures. The key factors have been a lack of political will, compounded by inconsistencies in both policy and legislation. Most of the policies seem to promote a sectoral approach to NRM. This has in a wider sense resulted in varying degree of policy overlap and conflict. The various pieces of legislation also lack provisions for conflict resolution and disputes among competing land users. Although the majority of national governments are promoting Community-Based Natural Resource Management (CBNRM) approaches, they lack the much needed commitment to provide the right enabling environment and technical support.

3.1.2. Institutional Setting

Key government institutions charged with the responsibility of managing forest resources in Miombo eco-region face a number of daunting challenges to effectively halt deforestation. The common challenges are poor staffing levels, inadequate funding to forestry operations, and lack of coordination with other line agencies. Government departments mandated to sustainably manage forest resources are poor staffed. Staffing levels are very low in most Forestry Departments in the region (Vinya *el al.* 2012a). Compounded by poor government funding to most Forestry agencies, forest utilization monitoring has often been relegated to a tertiary function in most countries in the Miombo eco-region. Funding to Forestry Departments in the Miombo eco-region has been nose-diving in the last two decades. Although Forestry Headquarters are regularly funded, the funding rarely gets down to the district forestry offices.

This state of affairs in the forestry sector in the Miombo eco-region has only worked to accelerate deforestation. Additionally, most institutions established by various pieces of legislation governing land and natural resources management tend to operate independently and in most cases with no defined coordination.

3.1.3. Political Pressure

During colonial years, a number of forest reserves were set aside for biodiversity conservation in all the miombo eco-region countries. These reserves were set aside with very good intentions and were gazette by existing pieces of legislation. The establishment of Protected Forest Areas (PFA) was meant to serve both national and local level forest needs. In most cases these PFA areas were set aside to provide national timber, protect water catchment areas, and general provision of ecosystem services. By statute, it is illegal to settle or extract timber from any PFA without a government permit.

However, increasing human populations in the miombo eco-region has generated immerse pressure for land. Not surprising some communities have found it easier to encroach onto PFA without any express permission from the relevant government agency. While the Forestry Departments have in some cases tried to remove illegal squatters from the Protected Forest Areas, it has in all instances faced resistance from squatter communities. Sad enough squatter communities have also sought protection from political leaders. For example in Zambia, mounting political pressure, has forced Forestry Department to degazett some of the heavily encroached PFA. An estimated 23,334 ha of PFA was degazetted in the period covering 2004 - 2008 to pave way for human settlement (Vinya *et al.* 2012a). This represents about 5,834 ha of PFA being lost or cleared of forest cover annually.

The problems of shortage of land for the resource-poor communities has been driven by poor land allocation which has seen large chunks of land being sold to a few foreign investors. This has pushing the most vulnerable members of community into landlessness condition. Therefore, their first line of defense is the nearest PFA which is rarely monitored by the District Forestry offices. Ideally, degazettion of PFA gives hope to other would be squatters that with patience any given piece of forest reserve can be excised. This in turn fuels the rising number of forest excisions that are seemingly on a rise in the miombo eco-region. Overall forest encroachments by squatters have resulted in permanent loss of forest cover.

3.2. Socio-Economic Factors

3.2.1. Deforestation and Poverty

Poverty is considered to be one of the key factors that undercut household ability to positively contribute to sustainable forest management in the Miombo eco-region. Rural and peri-urban areas are constantly faced with extreme poverty levels. These poverty levels have continued to fluctuate, due to economic instability in most countries in the Miombo eco-region. It is a well documented fact that high poverty levels are mainly driven by the national economic structures that have often failed to favour majority of the resource-poor rural folks.

Although most SADC governments have consistently expressed a wish to take development to rural areas, reality on the ground has demonstrated a negative trend. It is, therefore, not surprising that reliance on forest resources as a source of fall-back capital in all the Miombo eco-region countries has been increasing over the years.

3.2.2. Woodfuel Extraction

Miombo woodlands are the main source of fuelwood consumed in both rural and urban areas. The annual woodfuel consumption in the miombo eco-region is over 1,445 million m^3 (FAO 1998). This represents about 8% of the world consumption of wood fuel. In Zambia, for example, Kalumiana (1996) estimated that the per capita annual consumption of fuelwood is around 1,025 kg in rural areas and 240 kg for urban areas. This underlines the importance of wood fuel consumption as key driver in forest cover loss. Woodfuel is required for a number of reasons that include, charcoal production and firewood collection for curing fish, tobacco, and bricks.

Of all the woodfuel demand, charcoal production is by far the most dominant force behind temporal deforestation throughout the Miombo eco-region. The high energy demand in the urban areas provides a ready market for charcoal throughout the Miombo eco-region (Figure 2).

Fuelwood and charcoal production are estimated to contribute more than 3 percent of the countries' GDP, and account for more than 80 percent of the regional total energy household balance.

It is unarguable fact that Charcoal use has socio-economic benefits that accrue to numerous actors along the chain of custody (Malimbwi et al., 2005). The production, distribution and marketing employ over 15,000,000 people in the region (Kalinda et al., 2008).

a

b

Figure 2. a) A smouldering charcoal kiln in Miombo woodlands (Photo: Vinya J) b) Charcoal from Miombo woodlands being transported to urban markets in Zambia (Photo: Vinya R).

Interestingly, charcoal production tends to increase per capita income even when other sectors are not doing well in most rural areas. For example, the study of the contribution of charcoal to domestic income in Zambia showed that although rural per capita income from crop agriculture took a downward trend from US$37.07 in 1990 to US$17.33 in 2000, the contribution of charcoal increased from 65 percent to 83 percent over the same period (Chidumayo et al., 2001).

Tobacco curing is one of the main indirect drivers of fuelwood collection in Malawi, Zambia and Zimbabwe. The high international prices of tobacco are promoting increased production in the three countries. An additional driver of wood fuel extraction is the baking of clay bricks. The high regional price of cement is pushing the majority of resource-poor rural and urban population to embark on baked Kimberly bricks.

Furthermore, persistent poverty in the Miombo eco-region has often resulted in the dominance of charcoal use at domestic levels. It is assumed by most households that charcoal is a cheap source of domestic energy (Chiwele and Syampungani, 2011). Consequently, urbanization compounded with high poverty levels can be seen as an environmental problem in the Miombo eco-region. It is important to note that charcoal production when properly managed may not directly lead to permanent deforestation (Chidumayo, 1997).

3.2.3. Agriculture

Although most governments in the Miombo eco-region promote agricultural intensification, the practice on the ground has been that of extensification. Poor soil fertility coupled with unavailability of mineral fertilizers forces majority of resource-poor rural communities to rely on short rotation shifting cultivation (Figure 3). In Zambia, for example, small scale agriculture expansion makes a significant contribution to the loss of forest cover (Vinya et al. 2012a). Given that most domestic economies in the Miombo eco-region are agro-based, the emergence of agricultural expansion as a major driver of deforestation is not surprising. Agriculture provides the bulk of the regional food and cash requirements for the majority of the resource-poor rural communities. Therefore, large-scale agricultural systems combined with shifting cultivation have been the principal force behind forest cover loss in the Miombo eco-region.

In areas with high population density, Slash-and-burn and semi-shifting cultivation practices are hindering rapid forest regeneration process, thereby affecting the overall forest cover.

Figure 3. A newly opened up small scale agricultural field in Zambia (Photo: Vinya R).

Equally political upheavals in the last decade in Zimbabwe have accelarated human migrations to neighbouring countries in search of farming areas. This state of affairs is leading to opening up of forested land to pave way for crop production. Although the contribution of domestic animals grazing to the overall picture of deforestation is slightly insignificant in the Miombo eco-region, there is enough evidence to suggest that, at local level overgrazing is a factor in the whole deforestation equation. In Zambia, for example, cattle overgrazing is a leading contributor to forest cover loss in Southern and Western provinces (Vinya et al. 2012a). The problems of overgrazing are exacerbated by poor range land management practices that ignore the basic principles of rotational grazing (Caughley, 1975; Jachmann, 1994). The high concentration of wildlife animals for example, has resulted in vegetation damage with some of the areas being left devoid of woody vegetation.

3.3. Infrastructure Development

Majority of the countries in southern Africa are experiencing rapid urbanization coupled with industrialization. New housing units are being constructed to meet the ever growing urban population. In the dominantly mining economies such as Zambia and Congo DRC, huge expanses of forests are being cleared to pave way for large scale investments in the mining sector. Both mining and its related infrastructure development are linked to forest land clearance. Often, huge tracts of forested land are cleared to provide space for infrastructure development in the form of housing and roads construction. Furthermore, evidence has shown that infrastructure development is often followed by increased demand for construction timber. This state of affairs leads to an increase in exploitation pressure on both natural and plantation forests.

4. CHALLENGES FOR REDUCING DRIVERS OF DEFORESTATION

There are a number of daunting challenges that need to be addressed before deforestation and forest degradation in the Miombo eco-region can be effectively slowed down. The major challenges are enhancing synergies among relevant institutions dealing with NRM, updating national forests data bases, enhancing policy complementarity and de-politicizing forest governance.

4.1. Enhancing Synergies among Relevant Institutions Dealing with NRM

In many Miombo eco-region countries, forest management efforts are still undertaken in a sectoral and uncoordinated manner, with limited institutional enforcement and support. Enhancing synergies between the Forestry Departments and other sectors such as those of Energy and Agriculture has been a thorny issue to deal with. Unless these synergies are well strengthened it will be very difficult to attain sustainable forest management in most of the Miombo eco-region countries.

4.2. Updating National Forests Data Bases

Forest inventory data bases are out of date in the majority of countries in the Miombo eco-region. The existing information on the forests and woodlands in Miombo eco-region is outdated and in most cases incomplete.

Most of the countries face daunting challenges in updating their forest inventory data bases. The last extensive forest inventories in the majority of the countries in the Miombo eco-region were last undertaken in the 1960s. Although some funding for large scale forest inventories has been made available to the Miombo eco-region countries, the inventory design and sampling intensity do not meet national needs. A major challenge, therefore, lies in updating the existing forest databases. Though in some countries there maybe need to establish new databases altogether.

4.3. Enhancing Policy Complementarity

In the majority of the countries in the Miombo eco-region forest policies are in most cases at variance with other policies dealing with natural resources management. This has to a very large extent heightened conflict in land-use practices. In most cases, forestry has fallen victim to other land-uses such as mining and agriculture. The major challenge is, therefore, to promote synergies between forest policies and other policies that impact negatively on the forest environment such as agriculture.

4.4. De-Politicizing Forest Governance

Political pressure appears to be one of the main drivers of forest cover loss in the miombo eco-region. There is, therefore, a compelling need to minimize political pressure on the management of forest resources in the miombo eco-region. In order to overcome this challenge there is need to raise the level of awareness among all political players on the importance of forests to the overall national and regional development.

5. Mitigation Measures to Reduce Impacts of Deforestation in Miombo Woodlands

Addressing deforestation in the miombo eco-region presents two complementary strategies for mitigating climate change. First, by reducing forest deforestation, it is expected that greenhouse gas (GHG) emissions associated with forest cover loss can be minimized (Syampungani and Chirwa,

2011). Second, practicing sustainable land management can lead to carbon sequestration. Given that woodlands carbon loss is closely associated with loss of vegetation cover and soil erosion, mitigation measures that slow or reverse these processes are lead to increased forest carbon stocks (Trumper et al., 2008). A number of forest-based and non-forest based mitigation strategies that increase carbon stocks in degraded sites include; management of woodland regeneration (Chazdon 2008), sustainable agriculture practices (Nair et al. 2009), intensive management of woodlands fires and management of the energy sector.

5.1. Management of Woodlands Regeneration

Miombo woodland re-growth stands are highly productive ecosystems with high photosynthetic rates (Geldenhuys, 2005). Re-growth miombo woodlands like any other young forest stands have a high uptake of carbon dioxide. Following years of forest cover loss there has been an increase in natural regeneration.

These recovering degraded sites are thought to be sequestering significant quantities of carbon. In order to promote natural regeneration there is need to devise better forest management strategies that promote coppice management. This calls for adapting forest management practices that promote seedling survival in the wild. There may also be need to devise forest management strategies that aim at protecting new shoots against destructive late miombo fires, animal grazing or drought. Experimental studies have demonstrated that increasing stump height results in enhanced stump and coppice survival rates (Grundy, 1990). There is also evidence to suggest that felling methods have a significant effect on vigour and overall growth pattern during the initial stages of coppice development (Chidumayo 1993; Chidumayo et al., 1996).

5.2. Sustainable Agricultural Practices

Despite the prospects of improving the living standards of the rural poor under community based forest management arrangement, declining food production continues to poses a threat to sustainable forest management in the Miombo eco-region. Declining food production is closely linked with low soil fertility in sub-Saharan Africa and has in many cases compromised the domestic food security for the rural poor (FAO, 2001). The low soil fertility has its roots in low soil organic matter management (Bationo and Buerkert, 2001; Katyal, et al, 2001; Palm et al, 2001). This is linked to the wider poor soil management practices that are prevalent within Sub-Saharan Africa. However, the net result of poor soil fertility management has been the adoption of unsustainable farming practices (Barbier, 2000; FAO, 2001; Conelly and Chaiken, 2000; Kwesiga et al, 1999) by the wider resource-poor subsistence farmers as a short term copping strategy. However, adopting unsustainable farming practices often lead to the reduction of forest cover. Conversely, low-cost Agroforestry technologies not only do they offer the resource-poor farmers an opportunity to replenish their soil fertility, but also stand to reverse the negative relationship between farming and sustainable forest management. Many authors have demonstrated that Agroforestry offers a key option for dealing with problems of deforestation linked to poor farming practices and CO_2-induced global warming (Albrecht and Kandji 2003). In terms of carbon sequestration Agroforestry offers reliable solutions to mitigate the potential impacts of CO_2emmissions (Nair et al., 2009).

5.3. Intensive Fire Management Regimes

The role of fire in structuring Miombo woodlands has been emphasized by many authors (Bond et al. 2005; Chidumayo 1988; Koller et al. 2011; Mapaure and Campbell 2002). Many studies have shown that Miombo species respond individualistically to fires (Lawton 1972). On the basis of fire tolerance, Miombo woodlands tree species have been grouped into four main categories. These are: Fire-intolerant species: This group cannot survive fire and are known to occur where fires are excluded. Examples in this category include *Parinari excelsa, Entandophragma delevoyi* and *Syzygium guineense*. The second group is characterized as fire-tender species: These are species which have demonstrated population decline under regular burning and, therefore, increase under complete protection. This category is predominantly composed of the principal miombo species such as *Brachystegia spiciformis, Isoberlinia angolensis* and *Julbernadia paniculata*. The third group comprises semi-fire tolerant species: These species include among many *Maranthes polyandra, Uapaca kirkiana, Baphia bequeartii* and *Strychnos pungens*. These are relatively unaffected by early dry season fires but show signs of population reduction under late dry-season fires. The last group is the fire-tolerant or fire-hard species: Fire-tolerant species have the capability of surviving regular late dry-season bush fires. Examples in this category include *Pterocarpus angolensis*, *Parinari curatellifolia* and *Pericopsis angolensis*. Therefore, a successful fire management strategy should be one that reflects the varying fire tolerances of the Miombo woodland tree species. A better and well tried fire management regime that promotes growth of miombo species has been the use of early burning. In addition to early burning there are other practical steps that reduce fire risks in woodlands. These practices include piling of all the discarded wood debri away from stumps and patches of dense saplings followed by burning. Grazing also presents an opportunity in woodlands fire management as it maintains low grass levels. The reduction in the herbal layer as a result of grazing reduces woodlands fuel load and the damaging consequences of late bush fires.

5.4. Management of Domestic Energy

Fuelwood remains the primary source of domestic energy for the majority of resource-poor rural communities in miombo eco-region. The high level of fuelwood use is not surprising given low electrification rates in the SADC region (CPS 2008). This state of affairs can best be resolved by properly re-strategizing the regional energy sector. There are a number of available environmentally friendly energy options that the region can tap into. Bio-energy for example presents an alternative modern and more efficient use of biomass energy.

The SADC region is well endowed with huge reservoir of bio-energy material (plant materials, twigs, leaf litter, agricultural residues and dung) that may be exploited for Bio-energy production. Going through this route may demand for the strengthening of the linkage between the Forestry, Agriculture and Energy sectors.

Also minimizing the negative environmental impacts of charcoal production and consumption on woodland tree cover can be achieved by using improved and efficient stoves (Ezatti et al. 2000). There is ample evidence suggesting that improved charcoal stoves have high combustion and heat transfer efficiency than the traditional ones (Muneera and Mohamed 2003). Overcoming these challenges requires supportive policy environment and good governance. In order to foster adaptation of alternative energy sources, there is a need to develop programmes that would provide assistance to rural vulnerable communities.

6. Woodland Recovery, Carbon Stocks and Potential for Carbon Trading

The potential for carbon trading in the Miombo eco-region are substantial. Available above-ground standing biomass is estimated to be around 32,537 million tonnes (FAO, 2009). Therefore, the existence of huge expanses of forested land offers the potential for this eco-region to effectively participate in global carbon trade. Furthermore, deforestation that is not accompanied by infrastructure development presents yet another opportunity through which the region could benefit from carbon markets, through management of regeneration and rehabilitation of degraded forest lands.

6.1. Woodland Recovery

Africa's Miombo woodlands have time and again demonstrated high levels of resilience to vegetation disturbance as it regenerates virtually unchanged following clearing through agriculture or charcoal production (Syampungani and Chirwa, 2011). This ability to recovery has often been linked to the rooting habits of most Miombo woodlands species (Grundy, 1995; Chidumayo, 1985; Mistry, 2000).

It is, therefore, not surprising that average growth rate of between 4.4 and 5.6 mm has been observed in re-growth stands, compared to 2.3 and 4.8 mm in uneven aged mature stands. The re-growth stands are said to be part of the highly productive ecosystems.

These highly productive ecosystems tend to have high rates of photosynthetic processes and hence a high uptake of carbon dioxide. It therefore, follows that even with high levels of deforestation in Zambia degraded sites may still be managed for carbon sequestration through coppice or regeneration management.

6.2. Carbon Stock Levels and Potential for Carbon Trading

Standing forests in Miombo eco-region hold a considerable amount of total above-ground biomass. FAO (2009), estimate that the Miombo eco-region holds an estimated 16,268 million tons of above-ground carbon. These carbon stocks vary considerably across the region with DR Congo and Angola holding more than 50% of the regional carbon stocks (Figure 4; excludes carbon stocks in rainforests).

Malawi appears to hold the least quantities of carbon stocks in the Miombo eco-region. This maybe directly attributed to the high rate of deforestation that the country has experienced in the last three decades. The low carbon levels in Malawi, Mozambique and Zimbabwe also go to demonstrate the negative relationship between changes in land cover and carbon stock levels in any vegetation type. Comparative studies have shown that degraded sites contain 40 percent less carbon than well managed natural woodlands (Walker and Desanker, 2004). This clearly demonstrates that degraded Miombo woodlands sites are unsaturated with carbon and, therefore, have high potential to sequester carbon (FAO, 2004; Lal, 2004).

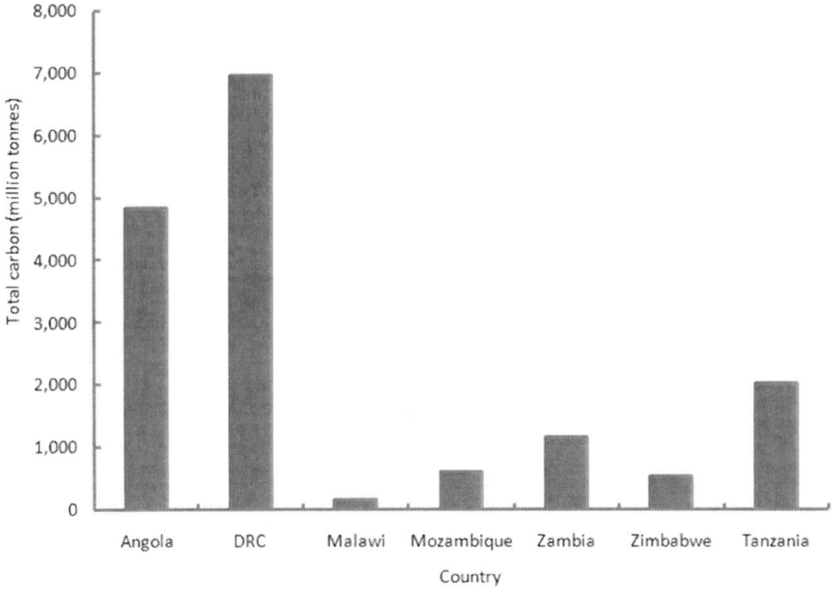

Source: FAO (2009).

Figure 4. Standing carbon stocks in the miombo woodlands by country.

REFERENCES

Albrecht A and Kandji ST (2003) Carbon sequestration in tropical agroforestry systems. *Agriculture Ecosystem and Environment* 99: 15–27.

Barbier BE (2000) The economic linkages between rural poverty and land degradation: Some evidence from Africa. *Agriculture Ecosystem and Environment* 82: 355 – 370.

Bationo A and Buerkert A (2001) Soil Organic carbon management for sustainable land use in Sudano-Sahelian west Africa. *Nutrient cycling in Agroecosystems* 61:131 – 142.

Bond WJ, Woodward FI and Midgley GF (2005) The global distribution of ecosystems in a world without fire. *New Phytologist* 165: 525–538.

Campbell B, Frost P and Byron N (1996) Miombo woodlands and their use: overview and key issues. In: B. Campbell (Ed.), *The Miombo in Transition: Woodlands andWelfare in Africa*, (Bogor: Center for International Forestry Research) pp. 1–10.

Chazdon RL (2008) Beyond Deforestation: Restoring Forests and Ecosystem Services on Degraded Lands. *Science* 320 (5882): 1458-1460.

Chidumayo EN (1988) Early post-felling response of Marquesia woodland to burning in Zambia Copperbelt. *Journal of tropical Ecology* 4, 361-372.

Chidumayo EN (1993) *Responses of Miombo woodland to Harvesting: Ecology and Management.* Stockholm Environment Institute/IT Publications, London.

Chidumayo EN (1997) *Miombo ecology and management: An introduction.* Stockholm Environment Institute, Stockholm.

Chidumayo EN, and Frost EN (1996) Population biology of Miombo trees. In: Campbell, B (Ed.), *The Miombo in Transition: Woodland and Welfare in Africa*. CIFOR, Bogor pp. 56-71.

Conelly WT and Chaiken SM (2000) Intensive farming, agro-diversity, and food security under conditions of extreme population pressure in Western Kenya. *Human Ecology*. 28 (1):19 – 51.

Ezzati M, Mbinda BM and Kammen DM (2000) Comparison of Emissions and Residential Exposure from Traditional and Improved Cookstoves in Kenya. *Environmental Science and Technology* 34 (4): 578–583.

FAO (2001) *Soil fertility management in support of food security in sub-Saharan Africa*. FAO. Rome, Italy.

FAO (2005). *Global Outlook Study for Africa. Sub regional Report: Southern Africa*. African Development Bank/European Commission/FAO, Rome, Italy.

Fearnside PM and Laurance WF (2003) Tropical deforestation and greenhouse-gas emissions. *Ecological Applications* 14:982–986.

Franzel S, Place F, Phiri D and Simwanza CP (1999) *Sesbania sesban* improved fallows in eastern Zambia: Their inception, development and farmer enthusiasm. *Agroforestry Systems* 47: 49 – 66.

Grundy IM (1990) *The potential for management of indigenous woodlands in communal farming areas of Zimbabwe with reference to the regeneration of Brachystegia spiciformis and Julbernadia globiflora.* Msc. Thesis, University of Zimbabwe, Harare.

Henry M, Maniatis D, Gitz V, Huberman D and Valentini R (2011) Implementation of REDD+ in sub-Saharan Africa: state of knowledge, challenges and opportunities. *Environment and Development Economics* 16: 381-404.

Katyal JC, Rao NH and Reddy MN (2001) Critical aspects of organic matter management in the tropics: The example of India. *Nutrient cycling in Agroecosystems* 61: 77 – 88.

Koller R and Samimi C (2011) Deforestation in the Miombo woodlands: a pixel-based semi-automated change detection method. *International Journal of Remote Sensing* 32 (22): 7631–7649.

Kutsch WL, Merbold L, Ziegler W, Mukelabai MM, Muchinda M, Kolle O and Scholes RJ (2011) The charcoal trap: Miombo forests and the energy needs of people. *Carbon Balance and Management* 6 (5):1-11.

Kwesiga FR and Coe R (1994) The effect of short rotation *Sesbania sesban* planted fallows on maize yield. *Forest Ecology and Management* 64: 199 – 208.

Lal R (2004). Soil Carbon Sequestration Impacts on Global Climate Change and Food Security. *Science* 304: 1623-1627.

Lusambo LP, Monela GC, and Katani J (2007) Socio-economic analysis of land use factors causing degradation and deforestation of Miombo Woodlands In Kilosa District, Tanzania. *Tanzania Journal of Forestry and Nature Conservation* 76: 28-39.

Malimbwi RE, Zahabu E, Monela GC, Misana S, Jambiya GC and Mchome B (2005) Charcoal potential of Miombo woodlands at Kitulangalo, Tanzania. *Journal of Tropical Forest Science* 17(2): 197-210.

Mapaure IN and Campbell BM (2002) Changes in miombo woodland cover in and around Sengwa Wildlife Research Area, Zimbabwe, in relation to elephants and fire. *African Journal of Ecology* 40 (3): 212–219.

Muneera SET and Mohamed EWM (2003) Adoption of biomass improved cook stoves in a patriarchal society: an example from Sudan. *Science of the Total Environment* 307: 259–266.

Nair PKR, Kumar BM, and Nair VD (2009) Agroforestry as a strategy for carbon sequestration. *Journal of Plant Nutrition and Soil Science* 172: 10–23.

Palm CA, Giller KE, Mafongoya PL and Swift MJ (2001). Management of organic matter in the tropics: Translating theory into practice. *Nutrient Cycling in Agroecosystems* 61: 63 – 75.

Ryan C (2009) *Carbon Cycling, Fire and Phenology in a Tropical Savanna Woodland in Nhambita, Mozambique*. PhD Thesis. University of Edinburgh, United Kingdom.

Stromgaard P (1986) Early secondary succession on abandoned shifting cultivator plots in the Miombo of South Central Africa. *Biotropica* 18 (2), 97-106.

Syampungani S, Chirwa PW, Akinnifesi FK, Sileshi G, and Ajayi CO (2009). The miombo woodlands at the cross roads: Potential threats, sustainable livelihoods, policy gaps and challenges. *Natural Resources Forum* 33 (2): 150–159.

Syampungani S and Chirwa PW (2011). Miombo Woodland Productivity: The Potential Contribution to Carbon Sequestration and Payment for Environmental Services in East and Southern Africa. In: Wallace, E.B (ed.), *Woodlands: Ecology, Management and Conservation*. Nova Science Publishers, Inc., New York.

Vinya R, Syampungani S, Kasumu EC, Monde C, and Kasubika R (2012a). *Preliminary study on the drivers of deforestation and potential for REDD+ in Zambia. A consultancy report prepared for Forestry Department and FAO under the national UN-REDD+ Programme*. Ministry of lands and Natural Resources. Lusaka, Zambia.

Vinya R, Syampungani S, Kashole D, and Sitwala W, (2012b) *Forest Sector in Zambia: Progress, Challenges and Opportunities*. EFZ, Lusaka, Zambia.

Walker SM and Desanker PV (2004). The impact of land use on soil carbon in miombo woodlands in Malawi. *Forest Ecology and Management* 203: 345-360.

In: Woodlands
Editor: Lazaro Manzanares
ISBN: 978-1-62618-177-9

Chapter 5

SOIL MICROBIAL DIVERSITY IN DRY WOODLANDS OF CENTRAL-WESTERN ARGENTINA: RESPONSE TO ECOLOGICAL CONDITIONS

L. Noe* and A. Abril
Departamento de Recursos Naturales, Facultad de Ciencias Agropecuarias, Universidad Nacional De Córdoba, Córdoba, Argentina

ABSTRACT

The dry woodlands cover a great area of central-western of Argentina, characterized by an E-W precipitation gradient ranging from 500 to 100 mm. Accordingly, two arid eco-regions with different dry woodland type are defined: Arid Chaco and Monte. Considerable researches have been done on biodiversity comparison between both eco-regions focusing on animal and vegetation communities, whereas the microbial community diversity has been neglected. However, microorganisms play a key role in the organic matter decomposition and nutrient availability, so that the microbial study increases current understanding of the ecosystem functionality. We analized bacteria and

* L. Noe: Departamento de Recursos Naturales. Facultad de Ciencias Agropecuarias. Universidad Nacional de Córdoba. cc 509 – 5000 - Córdoba. Argentina. E-mail: lauranoe@agro.unc.edu.ar.

actinomycetes diversity of soil and litter in two dry woodlands of central-western Argentina and its response to ecological conditions, including seasonal climatic changes. The study sites were Chancaní Reserve in Arid Chaco woodland (31°22'S and 65°27'W) and Ñacuñan Reserve in Monte woodland (34º03'S and 67°58'W). At each site, three composite samples of soil (0-20 cm deep) and surface litter (0.16 m^2) were collected in wet and dry season. Genetic diversity by polymerase chain reaction-denaturing gradient gel electrophoresis (PCR-DGGE) of bacteria and actinomycete communities were performed and interpreted in terms of ribotype richness and frequency and similarity index. The DGGE profile analysis reveals 39 bacteria ribotypes and 31 actinomyceteribotypes, among them 20.5% and 29% respectively, were observed only once. Two actinomyceteribotypes showed 100% of frequency (present in all samples). The ribotype richness of bacteria and actinomycetes were similar in all samples, but actinomycete richness was different between soil and litter (16.75 vs. 9.75, respectively). The low similarity index (<0.80) in bacteria and actinomycete communities indicate a marked microbial diversity in dry woodlands of central-western Argentina. We concluded that a) there are great bacteria and actinomycete diversity in dry woodlands of central-western Argentina, and b) the structure of microbial community responds to ecological conditions. To our knowledge, this is the first study that analizes bacteria and actinomycte diversity in a comparative way between dry woodland from realistic ecological conditions.

INTRODUCTION

Arid ecosystemsare characterizedby an extremeclimatic variability (daily, annual and interannual) and a topographicandedaphic heterogeneity, resulting in a complex structure oflandscapes(Noy-Meir 1973). The central-western arid region of Argentina is an extensive dry area with an E-W precipitation gradient ranging from 500 to 100 mm. Accordingly, two eco-region with different dry woodland types are defined: Arid Chaco and Monte (Bucher 1982; Cabido et al. 1993; Iglesias et al. 2011).

Both eco-regions (Arid Chaco and Monte) are different by ecological conditions, which include: vegetation structure and climatic seasonality. Vegetation in Monte eco-region is lower in terms of density and biodiversity (Bucher 1982; Cabido et al. 1993; Iglesias et al. 2011) and changes in the dominance of woody species (*Larrea*spp dominance) in relations to Arid Chaco eco-region, (ClaverandRoig-Juñent 2001). On the other hand, Arid

Chaco has greater climatic seasonality (7 months without precipitation) than Monte.

Considerable information is available conserning the both eco-region since more than 20 years with the aim of expading knowledge about animal and plant communities of Monte and Arid Chaco (Carranza et al. 1992; Marone et al. 2000; ClaverandRoig-Juñent 2001; Pol andLópez de Casenave 2004; Barchuk et al. 2008; PirkandLópez de Casenave 2011; Serbent et al. 2011), whereas the microbial community diversity has been neglected. However microbial diversity is a key factor to increases understanding of the ecosystem functionality (Wardle et al. 2004; Paul 2007).

Microorganisms are highly diverse and widely spread (Nannipieri et al. 2003; Paul 2007; Hirsch et al. 2010, Chaudhry et al. 2012). They play a important role in the decomposition of organic matter and nutrient availability (Clegg et al. 2003; Wardle et al. 2004; Torres et al. 2005; Aneja et al. 2006; Paul 2007; Chaudhry et al. 2012). Several biotic and/or abiotic factors lead to the alteration of microbial community structure and composition, which may directly or indirectly influence soil functionality (Nannipieri et al. 2003; Young and Crawford 2004). Climatic conditions change the relationships between microbial groups. For example, Okoro et al. (2009) mention that in arid environments the major microbial groups are actinomycetes due to it high resistence to dry conditions.

In recent years, the evaluation of the microbial community has employed biomolecular approaches based on direct total DNA extraction from a substrate (metagenome), without the isolation of individual organisms (Lynch et al. 2004; Hirsch et al. 2010). Methods ofmetagenomicstudy includes polymerase chain reaction-denaturing gradient gel electrophoresis (PCR-DGGE) which is a powerful genetic methodology that has led to revolutionary advances in microbial diversity studies (Muyzer et al. 1993; Nannipieri et al. 2003; Lynch et al. 2004; Ascher et al. 2010). This methodology allows the assessment of microbial diversity in complex communities and to monitor their response to environmental changes (Muyzer et al. 1993; Nakatsu 2007; Hayden et al. 2010).

The majority of microbial diversity studies with metagenomicbiomoleculartechniches are mainly conducted in soil, while information of litter microbial communities is little known (Kubartová et al. 2009; Pascault et al. 2010). In dry woodlands litter is scare (by low primary productivity) and it does not acumulate superficially due to its fast decomposition by microorganisms during wet season, in adittion to animal remove and wind dispersal. In conquence, litter in arid soil do not form an organic layer on the bulk soil (Bucher et al. 2003, Torres et al. 2005).

We analized bacteria and actinomycete diversity of soil and litter in two dry woodlands of central-western Argentina and its response to ecological conditions including seasonal climatic changes.

MATERIALS AND METHODS

Study Site

The study was conducted in two dry woodland types of western-central Argentina located in Arid Chaco and Monte eco-regions (Figure 1).

Dry woodland from Arid Chaco have a tree layer dominated by *Aspidospermaquebracho-blanco* and, in a lesser degree, by *Prosopis*spp, with an abundant shrub layer of *Larreadivaricata*, *Mimozyganthuscarinatus* and *Acacia furcatispina*. Grasses (genera: *Trichloris*, *Gouinia*, *Setaria* and *Pappophorum*) are present in sites with low woody cover. Mean annual rainfall range 350-500 mm, concentrated during the summer.

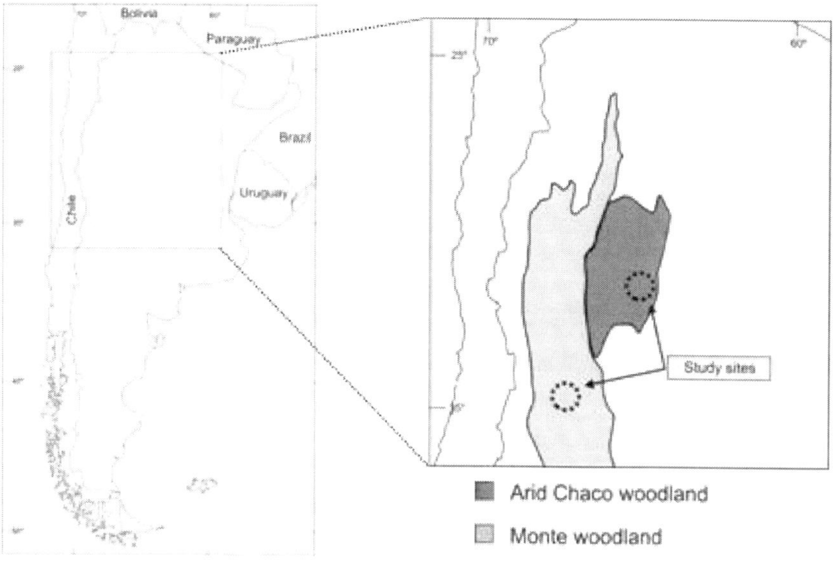

Figure 1. Study sites in Arid Chaco and Monte woodlands in central-western of Argentina.

In the dry winter season, water balance is negative resulting in a soil moisture deficit. Mean annual temperature is 20ºC (Carranza et al. 1992; Cabido et al. 1993). The soil is of alluvial origin, classified as MolicUltifluvent, with neutral pH, 25 mg g^{-1} of organic matter and 2.0 mg g^{-1} of total N (Abril et al. 2005; Noe 2012).

Dry woodland from Monte is an extensive shrubland dominated by *Larrea*spp, interspersed with open woodland of *Prosopisflexuosa*. The herb layer is composed by grasses, mainly perennial *Poaceae* C4 species (genera *Pappophorus*, *Digitaria*, *Trichloris*, *Aristida* and *Sporobolus*). Mean annual rainfall range 80-350 mm, with a strong water deficit along year. Scarce and unpredictable rains could fall in all months. Mean annual temperature range 13-15ºC (Claverand Roig-Juñent 2001).

The soils are TypicTorrifluvent, sandy-loam of aeolian and alluvial origin, neutral to slightly alkali pH, 13 mg g^{-1} of organic matter and 2.4 mg g^{-1} total N (Abril et al. 2009; Noe 2012).

In each woodland type, one 10 ha-sites were selected. In Arid Chaco woodland, the site was Chancaní Reserve (31° 22'S and 65° 26'W), whereas in Monte woodland, the site was Ñacuñán Reserve (34° 03'S and 67° 58'W). Both reserves were closed to wood extraction and grazing in 1980 and 1970, respectively.

Sampling Design

In each study site, three sampling 500 m-transects were randomly selected. In each transect, one composited sample (15 subsamples at random) of approximately 250 g of soil (0-20 cm) and one composited sample (15 subsamples at random) of litter in 0.16 m^2 (including leaf litterfall and woody material) were collected.

Samples were taken in two periods: dry and wet season, within the same week for all study sites.

Samples of two substrates (soil and litter) were stored refrigerated (4ºC) during transport, to avoid alterations in sample compositions. Soil samples were air dried for 24 h (25ºC), and sieved through a 2 mm.

Litter samples were air dried for 24 h (25ºC) and milled in laboratory steel mill (Lab Test, HBC, 23,000 rpm, ¼ HP) for 10-15 s. The 3-samples (45 sub-samples) from each situation were pooled together and stored at -20°C until DNA extraction.

DNA Extraction and PCR-DGGE Analysis

Total DNA was extracted from 0.5 g of soil and 0.3 g of litter (Fast DNA Kit for Soil BIO 101, Qbiogen) and quali-quantitatively characterized by agarose gel electrophoresis (purity and molecular weight) and spectrophotometer (Picodrop) in terms of yields (μg DNA g^{-1} soil/ litter) (Ascher et al. 2010). Bacteria and actinomycetes inhabiting soil and litter were screened by polymerase chain reaction-denaturing gradient gel electrophoresis (PCR-DGGE).

Bacteria community was assessed by 16S rRNA gene fragment PCR (16S rDNA) with 40 ng of DNA template, using the primer set GC-968f/1401r, generating 473 bpamplicons (Nübel et al. 1996). Actinomycetes were assessed by 16S rDNA PCR, performing a half nested PCR-approach (Heuer et al., 1997) with 40 ng target DNA using the primer set 243f/1401r (1st round PCR; 1175 bp) and GC- 968f/1401r (2nd round PCR, 473 bp) (Heuer et al. 1997; Clegg et al. 2003). PCR products were checked for correct size and estimation of quantity (for DGGE analysis) by gel electrophoresis in comparison to standardized DNA marker (Mass Ladder DNA Mix, Fermentas).

The DGGE of bacteria and actinomycetes communities were performed with 100 ngamplicons on 6% polyacrylamide at 60°C for 16 h, at 100 V, and 45-65% denaturing gradient (100 % denaturant contains 7 M urea and 40 % formamide). DGGE were performed in duplicates, stained by SybrGreen and the total number of different bands was determinated for the compared samples (Ceccherini et al. 2008). A standardized DNA (Mass Ladder DNA Mix, Fermentas) was used as intra-gel DGGE marker. PCR and DGGE were performed using Thermocycler (My-Cycler, Biorad) and IngenyPhorU system (Ingeny, Leiden, NL), respectively.

Calculations and Statically Analysis

The DGGE were interpreted in terms of ribotype richness (number of prominent DGGE bands) (Ascher et al. 2010). A number was designated for each band (#).

The similarity of the PCR-DGGE profiles among samples was assessed by similarity index (Dice Coefficient) according to the following relationship Cd= 2 a/(b+c) where a is the number of bands share by two samples, b is the number of bands in sample 1, and c is the number of bands in sample 2

(Clegget al. 2003). An index value ≥0.80 was considered that two communities have a similar genetic diversity.

Differences in DNA yield, and ribotype richness between woodland types, substrate types (soil and litter) and season (dry and wet) were tested using ANOVA and comparison of means by LSD test ($p \leq 0.05$). Data analyses were performed using the InfoStat (2001) software.

RESULTS

DNA Yield

The extracted total DNA from soil and litter samples from all study situations showed high molecular weight. Total average and range of DNA yield were 29.79 $\mu g\ g^{-1}$ (9.40-73.23 $\mu g\ g^{-1}$).

The DNA yield was significantly higher in Arid Chaco (40.95 $\mu g\ g^{-1}$) than in Monte (25.57 $\mu g\ g^{-1}$; p= 0.0001), in litter (40.95 $\mu g\ g^{-1}$) than in soil (18.64 $\mu g\ g^{-1}$; p= 0.0021), and in wet season (46.61 $\mu g\ g^{-1}$) than in dry season (12.98 $\mu g\ g^{-1}$; p= 0.0001).

Bacteria Community

Thirty-nine ribotypes were identified from the study situations: 35 in Monte and 34 in Arid Chaco. Total average and range of ribotype richness were 16.5 and 7-21, respectively. In the Monte, rybotype average was 14.46 (range 7-21) and in Arid Chaco 18.25 (range 13-21). Ribotype richness did not show differencies between woodland and substrate types and seasons.

No ribotype was represented in all situations. Five ribotypes were uniques of Monte and four ribotypes were uniques of Arid Chaco, whereas eight ribotypes were presented in one situation only (12.5% frecuency). Ribotypes that presented the most hight frequency (82.5%) were the #6 and #7 ribotypes (Figure 2a). None of the possible pairwise comparisons (28) exceed 0.80 of similarity index (Figure 3a). The highest similarity index was between the situations: soil of Montc in dry scason/soil of Arid Chaco in wct scason (0.78) and the lowest between soil of Monte in wet season/ litter of Monte in dry season (0.17).

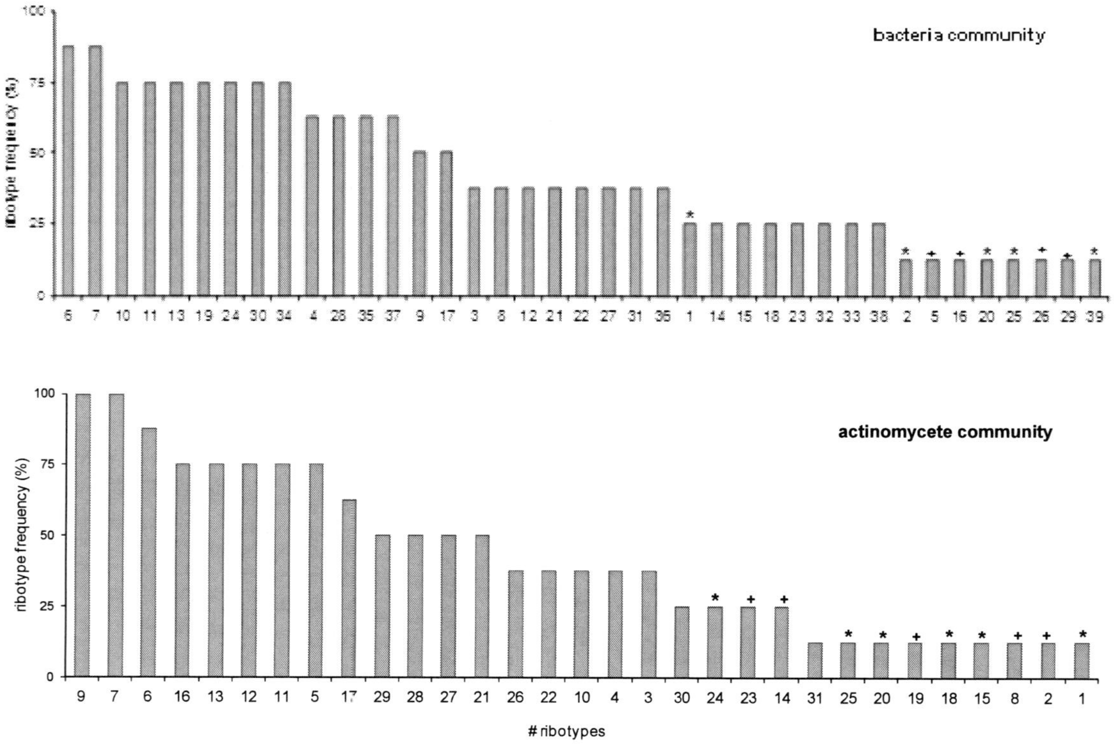

Figure 2. Bacteria and actinomycetesribotype frequencies (%) detected by PCR-DGGE analysis. + unique ribotypes from Arid Chaco woodland; * unique ribotypes from Monte woodland.

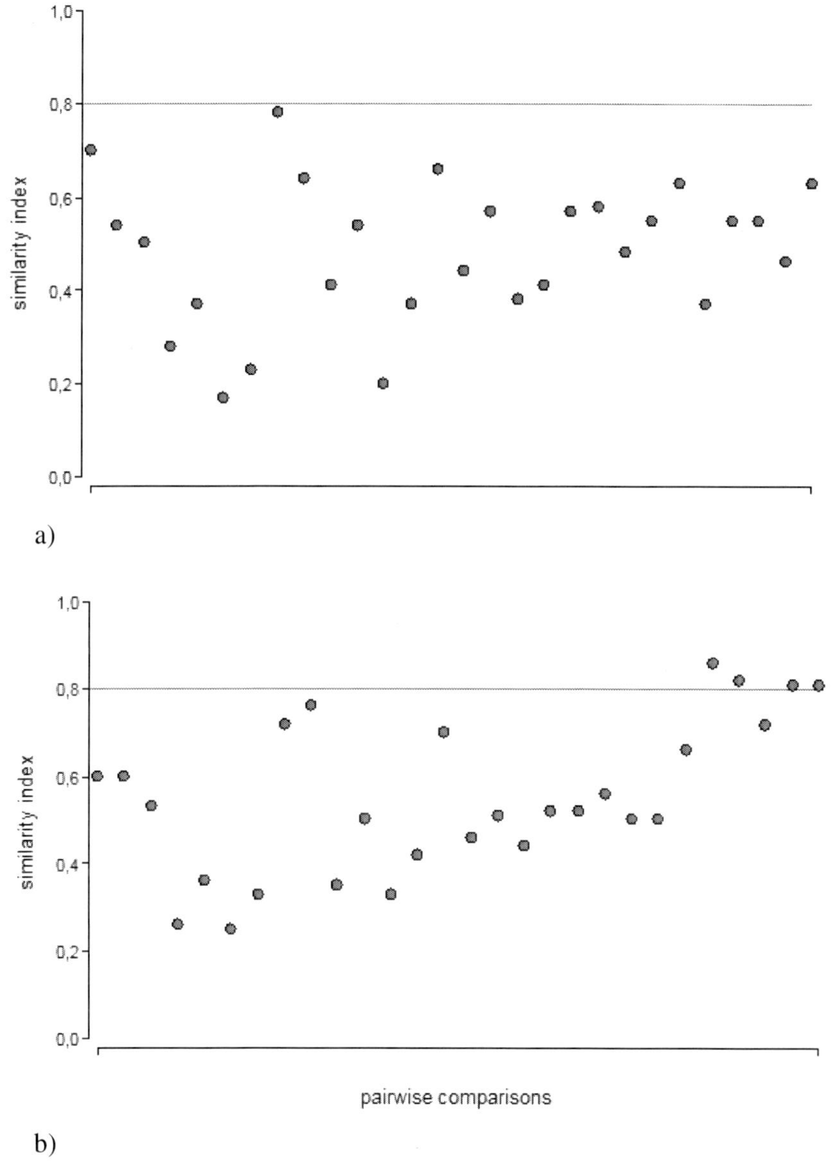

Figure 3. Genetic similarity indexes (pairwise comparison by Dice Coefficient) between all analyzed situations. (a) bacteria community, and (b) actinomycete community. The 0.8 line indicates similarity thershold.

Actinomycete Community

Thirty-one ribotypes were identified from study situations: 25 in Monte and 24 in Arid Chaco. Total average and range of ribotype richness were 13.25 and 8-19, respectively. The ribotype average in Monte was 12.5 (range 8-18) and in Arid Chaco 14 (range 9-19). The actinomyceteribotype richness was significantly higher in soil than in litter (16.75 vs 9.75, p=0.0124), while no differences were detected between woodland types and seasons.

Two ribotypes (#7 and #9) were represented in all situations (100% frecuency). Six ribotypes were uniques of Monte and six in Arid Chaco, whereas nine ribotypes were presented in one situation only (12.5% frecuency), (Figure 2b). Of the possible pairwise comparisons (28), only four pairs exceed 0.80 of the similarity index (Figure 3b). The highest similarity index was presented in Monte litter between dry and wet season (0.86) and the lowest between Monte soil in wet season/ Monte litter in dry season (0.17).

CONCLUSION

Ribotype Richness

Eveness of ribotype richness results are in agreement with Campbell et al. (2009), Clark et al. (2009) and Bachar et al. (2010) report that an equal ribotype amount of microbial communities in arid soils with different altitudes, vegetation and precipitations. Lynch et al. (2004) suggests that some minimum number of species is essential for ecosystem functioning under constat conditions, and that a large number of species is probably essential for maintaining the stability of ecosystem process in changing environments.

In comparison with Atacama desert, a major soil bacteria richness (39 vs14, respectively) and a lower actinomycetes richness (31 vs 46, respectively) was registered in our study (Okoro et al. 2009; Orlando et al. 2010). Hovewer, in our knowledge there is no information available on ribotype richness in litter, particularly in arid regions, which hinders the comparison of the data obtained in this study.

It is important to highlight that 24 % of the detected ribotypes of both communities (bacteria and actinomycete) correnpond to unique ribotypes of each situation.

It is mentioned that soil microbial diversity is enormously high, and most of the bacterial taxa are rare and may represent up to 37 % of total microbial load (Chaudhry et al. 2012). This is in agreement with Azua-Bustos et al. (2012) who reported that arid zone represent an effective "islands of biodiversity" due to the need to adapt to the extreme climatic conditions. In this context, microorganisms have great advantages due to their reproduction characteristics and their high generation rates which increase mutation probabilities (Van Gestel et al. 1993; Okoro et al. 2009; Schaechter 2009).

Schaechter (2009) and Azua-Bustos et al. (2012) reported that natural UV radiations induce high mutation rates in microorganisms, so that in arid environments with high insolation, UV radiation could justify the unique ribotypes. This is in agreement with novel taxa of simbiotic and no-simbiotic N_2 fixing bacteria in the same area (Velázquez et al. 2001; Rivas et al. 2004) and actinomycetes from Atacama desert soil (Okoro et al. 2009, Santhanam et al. 2012 a, b).

It is important to highlight the dominance of actinomycetes comparing to bacteria (relative richness) in soil, while litter present an inverse pattern (Figure 4). This observation accords with Boudjella et al. (2006), Chanal et al. (2006) and Okoro et al. (2009) who concluded that the major components of the arid soil microbial community are members of the class Actinobacteria.

The lower actinomyceteribotyperichenss in litter than soil could be explained by pH differences. Several authors mention a strong correlation between actinomicetes and pH (Hill et al. 2011; Paul 2007).

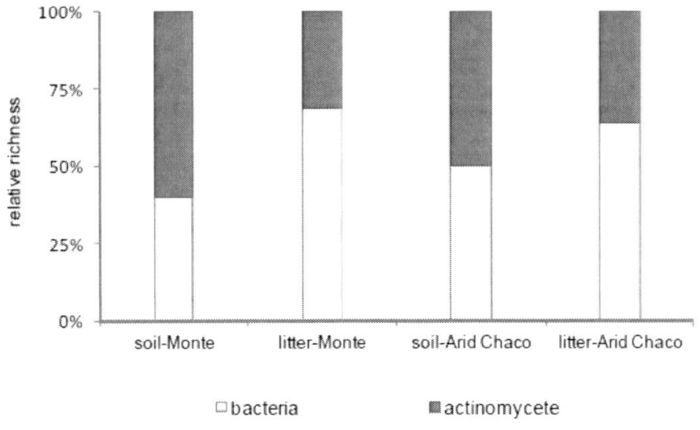

Figure 4. Bacteria and actinomycetes relative richness (%) in soil and litter from Arid Chaco and Monte woodland.

The actinomycetes prevail in neutral-alkaline soils which agree with soil pH values report by Noe (2012) (range 7.19 to 9.00) for the same study sites.

Otherwise, litter pH could be low due to organic acid production during litter decomposition (Paul 2007; Noe 2012).

Genetic Similarity

The scarce genetic similarity of the analized communities between woodland and substrate types (Figure 5) indicates the effect of ecological conditions on microbial diversity, showing that microbial community structure differs temporal and spatially, even in reduce scales (e.g. 6 months).

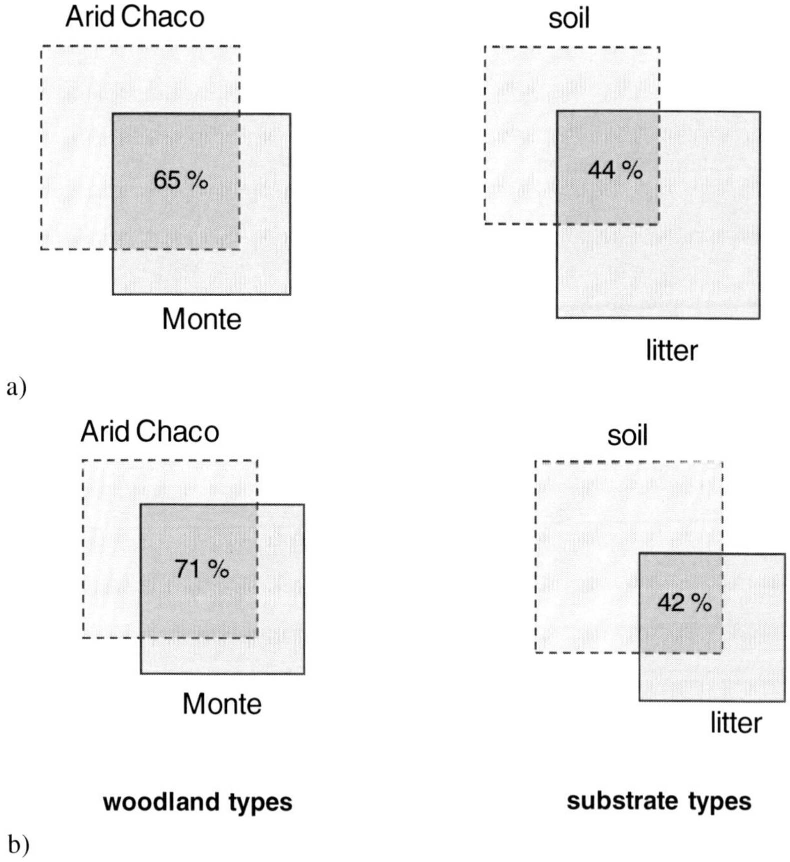

Figure 5. Shared ribotypes (%) between woodland and substrate types. (a) bacteria community, and (b) actinomycete community.

Several authors demostrated that soil microorganisms diversity is stronly affected by soil physical and chemical variables such as soil temperature, water content, pH, soil organic matter, available N, vegetation type, etc, which can have a more severe effect in desert ecosystems (Aneja et al. 2006; Clark et al. 2009; Kubartová et al. 2009; Ping et al. 2009; Orlando et al. 2010; Saul-Tcherkas and Steiberger 2011).

Differences in type and quantity of resources and physic-chemical conditions in a habitat (temperature, water, light, oxygen) define the niche for particular specie. The ecological theory indicates the existence of a principal niche for each species (Kubartová et al. 2009).

The very low similitude between the soil and litter microbial communities (Figure 5) is expected, due to the differences in chemical caracteristics of both substrates (Bossio et al. 2005; Meier et al. 2008). However, in contrast to the results mentioned by Saul-Tcherkasand Steinberger (2011), Noe (2012) did not found a relationship between litter and soil chemical characteristic, and genetic similarity dendrograms of microbial communities. This lack of relationships indicates that community structure can be controlled by factors other than the chemical characteristics (Bachar et al. 2010). Probably litter exposure to climatic conditions is a key factor. It is widely known the effect of insolation on microorganisms, which magnitud depends on the amount of litter acumulated (Schaechter 2009).

Another interesting question is the detection of two cases with similarity index in litter actinomycete community between woodland types. This could be due to the low number of taxa adapted to high insolation and starvation that it is founded on litter. It is improbable that spore dispersion by wind leads to this similarity like Azua-Bustos et al. (2012) reports, because woodlands are separated 370 km.

Seasonal Climatic Changes

The low bacteria community similarity between seasons (Figure 6a) agrees with the seasonal shifts in soil microbial communities observed by Gelsomino et al. (2006), who postulated that different climatic conditions change certain populations of the whole microbial community. Regarding to litter, changes in microbial communities could be resulted from microbial succession during decomposition process according to changes in chemical composition of litter (Torres et al. 2005; NoeandAbril 2008; Hossain et al. 2010; Carranza et al. 2012).

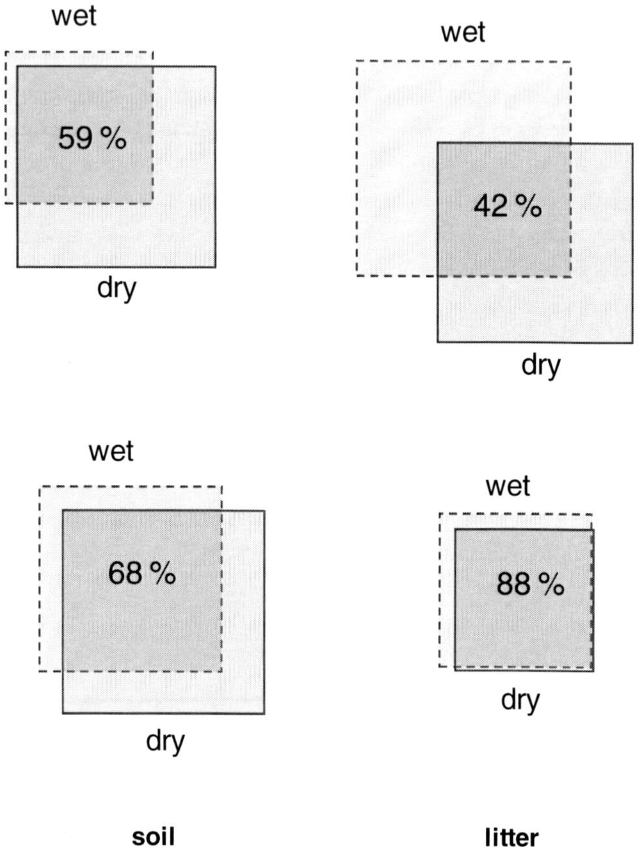

a)

b)

Figure 6. Shared ribotypes (%) between seasons in soil and litter. (a) bacteria community, and (b) actinomycete community.

High genetic similitude of litter actinomycete community between seasons (Figure 6b) could be explained by actinomycete resistance mechanisms. Spore formation allows the actinomycetes survival in substrate until better growth conditions of the next season (Van Gestel et al. 1993; Bell et al. 2009; Zenova et al. 2009; Pascault et al. 2010).

It is really difficult to explain the high actinomycete similitude index in litter between Arid Chaco-dry season and Monte-dry/wet season. This surprising result could be due a bias in DGGE methodology. It is mentioned that PCR-DGGE methods may preferentially target dominant bacterial types, in consequence, the specificity of the primer set used, may not be able to separate more complex soil microbial communities as actinomycetes (Clegg et al. 2003).

In summary: a) there are great bacteria and actinomycete diversity in dry woodlands of central-western Argentina, and b) the structure of microbial community responds to ecological conditions. To our knowledge, this is the first study that analizes bacteria and actinomycete diversity in a comparative way between dry woodland from realistic ecological conditions.

ACKNOWLEDGMENTS

This research was financially supported by SECyT, UNC to ABA. Noe L fellowship was supported by CONICET (National Research Council of Argentina).

REFERENCES

Abril, A., Barttfeld, P., Bucher, E. H. The effects of fire and overgrazing disturbs on soil carbon balance in the dry Chaco forest. *Forest Ecology and Management*, 2005 206, 399-405.

Abril, A., Villagra, P., Noe, L. Spatiotemporal heterogeneity of soil fertility in the Central Monte desert (Argentina). *Journal of Arid Environments*, 2009 73, 901-906.

Aneja, M. K., Sharma, S., Fleischmann, F., Stich, S., Heller, W.,Bahnweg, G., Munch, J. C., Schloter, M. Microbial colonization of beech and spruce litter-Influence of decomposition site and plant litter species on the diversity of microbial community. *Microbial Ecology*, 2006 52, 127-135.

Ascher, J., Ceccherini, M. T., Chronakova, A., Jirout, J., Borgogni, F., Elthottova, D., Simek, M., Pietramellara, G. Evaluation of the denaturing gradient gel electrophoresis-apparatus as a parameter influencing soil microbial community fingerprinting. *World Journal of Microbiology and Biotechnology*, 2010 26, 1721-1726.

Azua-Bustos, A., Urrejola, C., Vicuña, R. Life at the dry edge: Microorganisms of the Atacama Desert. *FEMS Letters*, 2012 586, 2939-2945.

Bachar, A., Al-Ashhab, A, Soares, M. I. M., Sklarz, M. Y., Gillor, O. Soil microbial abundance and diversity along a low precipitation gradient. *Microbial Ecology*, 2010 60, 453-461.

Barchuk, A., Iglesias, M. R., Boetto, M. N. Spatial association of *Aspidospermaquebracho-blanco* juveniles with shrubs and conspecific adults in the Arid Chaco, Argentina. *Austral Ecology*, 2008 33, 775-783.

Bell, C. W., Acosta-Martinez, V, McIntyre, N. E., Cox, S., Tissue, D. T., Zak, J. C. Linking microbial community structure and function to seasonal differences in soil moisture and temperature in a Chihuahuan desert grassland. *Microbial Ecology*, 2009 58, 827-842.

Bossio, D. A., Girvan, M. S., Verchot, L.,Bullimore, J., Borelli, T., Albrecht, A., Scow, K. M., Ball, A. S., Pretty, J. N., Osborn, A. M. Soil microbial community response to land use change in an agricultural landscape of western Kenya. *Microbial Ecology*, 2005 49, 50-62.

Boudjella, H.,Bouti, K., Zitouni, A., Matthieu, F., Labrihi, A., Sabauo, N. Taxonomic and chemical characterization of antibiotics of Streptosporangium Sg10 isolated from Saharan soil. *Microbiology Research,* 2006 16, 288-298.

Bucher, E. H. Chaco and Caatinga- South American arid savannas. Woodlands and thickets. In: Huntley, B. J., Walker, B. H., editors. *Ecology of Tropical Savannas*. Berlin, Alemania: Springer-Verlag; 1982; 48-79.

Bucher, E. H., Torres, P. A., Abril, A. B. Litter quality and litter removal by the native fauna in the western Chaco woodland of Argentina. *Journal of Tropical Ecology*, 2003 19, 337-341.

Cabido, M., González, C., Acosta, A., Díaz, S. Vegetation changes along a precipitation gradient in Central Argentina. *Vegetation*, 1993 109, 5-14.

Campbell, J. H., Clark, J. S., Zak, J. C. PCR-DGGE comparison of bacterial community structure in fresh and archived soils sample along a Chihuahua desert elevational gradient. *Microbial Ecology*, 2009 57, 261-266.

Carranza, C.,Noe, L., Merlo, C.,Ledesma, M.,Abril, A. Effect of forest clearing type on the decomposition of native and introduced pastures in the Arid Chaco, Argentina. *Revista de Investigaciones Agropecuarias*, 2012 38, 97-107.

Carranza, L., Cabido, M., Acosta, A., Páez, S. Comunidades vegetales del Parque Natural Provincial y Reserva Forestal Natural Chancaní, Provincia de Córdoba. *Lilloa*, 1992 38, 75-92.

Ceccherini, M. T.,Ascher, J.,Agnelli, A.,Certin, G.,Pietramellara, G.,Piovanelli, C.,Nannipieri, P. Tree bark and soil ammonia oxidizers: a molecular study on a historical forest of central Italy. *FreseniusEnvironmental Bulletin*, 2008 17, 882-889.

Chanal, A.,Chapon, V.,Benzerara, K.,Barakat, M., Christen, R., Achouak, W.,Barras, F.,Heulin, T. The desert of Tataouine: an extreme environment that hosts a wide diversity of microorganisms and radiotolerant bacteria. *Environmental Microbiology*, 2006 8, 514-525.

Chaudhry, V.,Rehman, A., Mishra, A.,Chauhan, P. S.,Nautiyal, C. S. Changes in bacterial community structure of agricultural land due to long-term organic and chemical amendments. *Microbial Ecology*, 2012 64, 450-460.

Clark, J., Campbell, J., Grizzle, H., Acosta, V., Zak, J. Soil microbial community response to drought and precipitaction variability in the Chihuahuan desert. *MicrobialEcology*, 2009 57, 248-260.

Claver, S., Roig-Juñent, S. A. El desierto del Monte: *La Reserva de Biosfera de Ñacuñán*. Mendoza. Córdoba, Argentina: Triunfar; 2001.

Clegg, C. D., Lovell, R. D. L., Hobbs, P. J. The impact of grassland management regime on the community structure of selected bacterial groups in soils. *FEMS Microbiology Ecology*, 2003 43, 263-270.

Gelsomino, A., Badalucco, L., Ambrosoli, R., Crecchio, C., Puglisi, E., Meli, S. M. Changes in chemical and biological soil properties as induced by anthropogenic disturbance: A case study of an agricultural soil under recurrent flooding by wastewaters. *Soil Biology and Biochemistry*, 2006 38, 2069-2080.

Hayden, H. L., Drake, J., Imhof, M., Oxley, A. P. A., Norng, S., Mele, P. M. The abundance of nitrogen cycle genes *amo*A and *nif*H depends on land-uses and soil types in South Eastern Australia. *Soil Biology and Biochemistry,* 2010 42, 1774-1783.

Heuer, H., Krsek, M., Baker, P., Smalla, K, Wellington, E. M. H. Analysis of actinomycete communities by specific amplification of genes encoding 16S rRNA and gel- electrophoretic separation in denaturing gradient. *Applied and Environmental Microbiology*, 1997 63, 3233-3241.

Hill, P., Kristufek, V., Dijkhuizen, L., Boddy, C., Kroetsch, D., van Elsas, J. D. Land use intensity controls actinobacterial community structure. *Microbial Ecology*, 2011 61, 286-302.

Hirsch, P. R., Mauchline, T. H., Clark, I. M. Culture-independent molecular techniques for soil microbial ecology. *Soil Biology and Biochemistry*, 2010 42, 878-887.

Hossain, M. Z., Okubo, A., Sugiyama, S. Effects of grassland species on decomposition of litter and soil microbial communities. *Ecological Research,* 2010 25, 255-261.

Iglesias, M. R., Barchuk, A., Grilli, M. P. Carbon storage, community structure and canopy cover: a comparison along a precipitation gradient. *ForestEcology and Management*, 2011 265, 218-229.

InfoStat. Universidad Nacional de Córdoba, Facultad de Ciencias Agropecuarias. 2001. URL:http://www.InfoStat.com.ar

Kubartová, A., Ranger, J., Berthelin, J., Beguiristain, T. Diversity and decomposing ability of saprophytic fungi from temperate forest litter. *Environmental Microbiology*, 2009 58, 98-107.

Lynch, J., Benedetti, A., Insam, H., Nuti, M., Smalla, K., Torsvik, V., Nannipieri, P. Microbial diversity in soil: ecological theories, the contribution of molecular techniques and the impact of transgenic plants and transgenic microorganisms. *Biology and FertilitySoils*, 2004 40, 363-385.

Marone, L., Horno, M. E., González del Solar, R. Post-dispersal fate of seeds in the Monte desert of Argentina: patterns of germination in successive wet and dry years. *Journal of Ecology*, 2000 88, 940-949.

Meier, C., Wehrli, B., van der Meer, J. R. Seasonal fluctuations of bacterial community diversity in agricultural soil and experimental validation by laboratory disturbance experiments. *Microbial Ecology*, 2008 56, 210-222.

Muyzer, G., De Waal, E. C., Uitterlinden, A. G. Profiling of complex microbial popilations by denaturating gradient gel electrophoresis analysis of polymerase chain reaction-amplified genes coding for 16S rRNA. *Applied and Environmental Microbiology*, 1993 59, 695-700.

Nakatsu, C. H. Soil microbial community analysis using denaturing gradient gel electrophoresis. *Soil Science Society of America Journal*, 2007 71, 562-571.

Nannipieri, P., Ascher, J., Ceccherini, M. T., Landi, L., Pietramellara, G., Renella, G. Microbial diversity and soil functions. *EuropeanJournal of SoilScience*, 2003 54, 655-670.

Noe, L., Abril, A. Interacción entre calidad de restos vegetales, descomposición y fertilidad del suelo en el desierto del Monte de Argentina. *Ecología Austral*, 2008 18, 181-193.

Noe, L. Efecto del sobrepastoreo sobre la diversidad funcional microbiana en zonas áridas de Argentina, Thesis, Facultad de Ciencias Exactas, Físicas y Naturales, Universidad Nacional de Córdoba. 2012.

Noy-Meir, I. Desert ecosystems: Environment and producers. *Annual Review of Ecology and Systematic*, 1973 4, 25- 41.

Nübel, U., Engelen, B., Felske, A., Snaidr, J., Wieshuber, A., Amann, R. I., Ludwig, W., Backhaus, H. Sequence heterogeneities of genes encoding 16S rRNAs in *Paenibacilluspolymyxa* detected by temperature gradient gel electrophoresis. *The Journal of Bacteriology*, 1996 178, 5636–5643.

Okoro, C. K., Brown, R., Jones, A. L., Andrews, B. A., Asenjo, J. A., Goodfellow, M., Bull, A. T. 2009. Diversity of culturableactinomycete in hyper-arid soils of the Atacama desert, Chile. *Antonie van Leeuwenhoek*, 2009 95, 121- 133.

Orlando, J., Alfaro, M., Bravo, L., Guevara, R., Carú, M. Bacterial diversity and occurrence of ammonia-oxidizing bacteria in the Atacama Desert soil during a "desert bloom" event. *Soil Biology and Biochemestry*, 2010 42, 1183-1188.

Pascault, N., Nicolardot, B., Bastian, F., Thiebeau, P., Ranjard, L., Maron, P. A. In situ dynamics and spatial heterogeneity of soil bacterial communities under different crop residue management. *Microbial Ecology*, 2010 60, 291-303.

Paul, E. *Soil Microbiology, Ecology and Biochemistry.* San Diego, EstadosUnidos: Academic Press. Inc.; 2007.

Ping, L. U.,Wanli, M. A., Hao, L. I., Yongsheng, W. U., Guifen, L. V., Jizheng, HE. Seasonal dynamics changes in bacterial compositions in the Inner Mongolia desert steppe. *Frontiers of Biology in China*, 2009 4, 187-191.

Pirk, G. I., Lopez de Casenave, J.Seed preferences of three harvester ants of the genus. *Annals of the Entomological Society of America,* 2011 104, 212-220.

Pol, R., Lopez de Casenave, J. Activity patterns of harvester ants *Pogonomyrmexpronotalis* and *Pogonomyrmexrastratus* in the Central Monte Desert, Argentina. *Journal of Insect Behavio*r, 2004 17, 647-661.

Rivas, R., Abril, A., Trujillo, M. E., Velázquez, E. *Sphingomonas-phyllosphaerae* sp. nov., from phyllosphere of *Acacia caven* in Argentina. *International Journal of Systematic and Evolutionary Microbiology*, 2004 54, 2147-2150.

Santhanam, R., Okoro, C. K., Rong, X., Huang, Y., Bull, A. T., Andrews, B. A.,Asenjo, J. A., Weon, H-Y.,Goodfellow, M. *Streptomyces deserti* sp. nov., isolated from hyper-arid Atacama Desert soil. *Antonie van Leeuwenhoek*, 2012a 101, 575–581.

Santhanam, R., Rong, X., Huang, Y., Andrews, B. A., Asenjo, J. A., Goodfellow, M.*Streptomyces bullii* sp. nov., isolated from a hyper-arid

Atacama Desert soil. *Antonie van Leeuwenhoek*, 2012b, DOI 10.1007/s10482-012-9816-x.

Saul-Tcherkas, V., Steinberger, Y. Soil microbial diversity in the vicinity of a Negev desert shrub *Reaumurianegevensis*. *Microbial Ecology*, 2011 61, 64-81.

Schaechter, M. *The Desk Enciclopedia of Microbiology*. San Diego, EstadosUnidos: Academic Press; 2009.

Serbent, M. P., Periago, M. E., Leynaud, G. C.*Mazamagouazoubira* (Cervidae) diet during the dry season in the arid Chaco of Córdoba (Argentina). Jo*urnal of Arid Environments*, 2011 75, 87-90.

Torres, P. A., Abril, A. B., Bucher, E. H. Microbial succession in litter decomposition in the Semi-arid Chaco woodland. *Soil Biology and Biochemistry,* 2005 37, 49-54.

Van Gestel, M., Merckx, R., Vlassak, K. Microbial biomass responses to soil drying and rewetting: the fate of fast and slow growing microorganisms in soils from different climates. *Soil Biology and Biochemistry*, 1993 25, 109-123.

Velázquez, E., Mariano Igual, M., Willems, A., Fernández, M. P., Muñoz, E., Mateos, P. F., Abril, A., Toro, N., Normand, P., Cervantes, E., Gillis, M., Martínez-Molina E. *Mesorhizobiumchacoense* sp. nov., a novel species that nodulates*Prosopis alba* in the Chaco Árido region (Argentina). *International Journal of Systematic and Evolutionary Microbiology*, 2001 51, 1011-1021.

Wardle, D. A., Bardgett, R. D., Klironomo, J. N., Setälä, H., van der Putten, W. H., Wall, D. H. Ecological linkages between aboveground and belowground biota. *Science*, 2004 304, 1629-1633.

Young, I. M., Crawford, J. W. Interactions and self-organization in the soil-microbe complex. *Science*, 2004 304, 1634-1637.

Zenova, G. M., Kurapova, A. I., Lysenko, A. M., Zvyagintsev, D. G. The structural-functional organization of thermotolerant complexes of Actynomicetes in desert and volcanic soils. *Soil Biology*, 2009 42, 531-535.

In: Woodlands
Editor: Lazaro Manzanares
ISBN: 978-1-62618-177-9

Chapter 6

DEVELOPING SILVICULTURAL SYSTEMS BASED ON DISTURBANCE-RECOVERY KNOWLEDGE OF THE SOUTHERN AFRICAN MIOMBO WOODLANDS

Stephen Syampungani[1*], Coert J. Geldenhuys[2] and Paxie W. Chirwa[2]

[1]School of Natural Resources, Copperbelt University, Kitwe, Zambia

[2]Faculty of Natural and Agricultural Sciences, University of Pretoria, Lynnwood Road, Pretoria, South Africa

ABSTRACT

The current negative perception that various land use systems within the Southern African Miombo woodlands cause the loss of Miombo woodland and degradation of the environment influenced the development of forest management programmes in the Miombo ecoregion. Such programmes exclude many forms of woodland utilization because the available silvicultural systems are not aligned with the ecological characteristics of Miombo woodland ecosystems or their targeted species. This chapter explores the concept of disturbance-recovery processes, with recovery development as the basis for

* Email: syampungani@cbu.ac.zm.

developing silvicultural systems suitable for managing Miombo woodland complexity. The size class profiles derived from various data sources are used to determine specific shade-or fire-tolerance characteristics of key economic and ecological tree species of Miombo woodlands. Additionally, the stand level grain analysis gives an indication of the scale of disturbances that drives Miombo woodland characteristics. Such information underlines the development and implementation of appropriate silvicultural management systems to ensure sustainable resource use from Miombo woodlands.

Keywords: Silvicultural management systems, disturbance, recovery

1. Introduction

Southern African woodlands support the livelihoods of over 100 million rural and urban dwellers through the provision of non-wood products which include bees wax, honey, edible fruits, edible insects, mushrooms and traditional medicines (Brigham et al. 1996). The major products from miombo woodlands are charcoal for rural and urban energy, water for downstream needs, timber and other wood products, grazing, hunting, fruit and other foodstuffs, honey, medicinal substances (Syampungani and Chirwa, 2011). They also provide the rural poor with the natural insurance or safety net for mitigating risks associated with the weather misfortunes like drought (Syampungani and Chirwa, 2011). There is therefore a very high level of forest dependence among the rural communities throughout the miombo ecoregion for both income generation and food security (Syampungani and Chirwa, 2011). This calls for sustainable management of the southern African woodlands. Sustainable woodland management is the process of managing permanent forest land to achieve one or more clearly specified objectives of management with regard to the production of a continuous flow of desired products and services without undue reduction of its inherent values and future productivity and without undue undesirable effects on physical and social environment (ITTO, 1992). It should therefore meet the needs of present societies without compromising the ability of future generations to meet their own needs (Geldenhuys, 1993).

Sustainable forest management systems are focussed on balancing the needs (rate of resource use) of users against the regeneration ecology and growth of the resource base to ensure the sustainable use and conservation of forest resources (Hartshorn, 1995). The need to understand the effects of harvesting on the composition and structure of residual stand is essential for developing optimum management systems (Cannon et al. 1995). Optimum management systems, although marginally alter the natural demography and biomass standing crop, provides forest use for a diverse range of uses and values. These include existence of a management plan, pre-logging and inventory, silvicultural prescriptions for major forest types, regulation of initial harvesting and subsequent cutting cycles and management of transition from first to second cutting cycles, presence of clear, official harvesting rules, monitoring of residual growing stock after logging, and extent of vegetation disturbance (Geldenhuys, 1993). Sustained uneven-aged management relies on the principle that as trees grow to the next diameter class, cutting of trees in excess of the desired number is observed. Theoretically, period yields should be regular and continuous if the selected diameter distribution allows regeneration of trees in residual stand and permits the established regeneration to develop into progressively larger size classes at a considerate rate, and if it is maintained by proper cutting (Geldenhuys, 1993). This requires developing a detailed understanding of the response of miombo woodlands to various disturbances, their frequency, magnitude and effect. It further calls for developing an interpretive understanding of specific size class profiles of both the individual key species and the woodland in general.

2. Size Class Structure and Implications for Woodland Management

Assessment of the stem diameter distributions of species across forest types may reveal either of these class profiles namely; *Inverse-J-shaped size class* or *Bell-shaped* or *static* shape (Figure 1). Each of these has different implications on developing sustainable management of species.

Inverse J-shaped size class (DWAF, 2005): The inverse *J*-shaped size class profile has a higher abundance of individuals in smaller size classes than in larger classes. This characteristic size class structure is indicative of a stable and expanding population (Geldenhuys, 1993). In the natural evergreen forest the shade-tolerant canopy species, such as *Podocarpus latifolius*, typically has this stem diameter because they continuously regenerate under the canopy (DWAF, 2005). However, in fire-adapted woodlands this type of curve is typical of the fire-tolerant species (Geldenhuys, 1993). According to Peter (2005), it is safe to assume that the death of an adult tree will, at some point, be replaced by individuals growing up from the smaller classes. This therefore implies that such a size class profile represents the ideal of a stable and self sustaining population (Mudekwe, 2006). It may be used for planning yield regulation for woodlands such as Miombo to ensure sustainable supply of raw material.

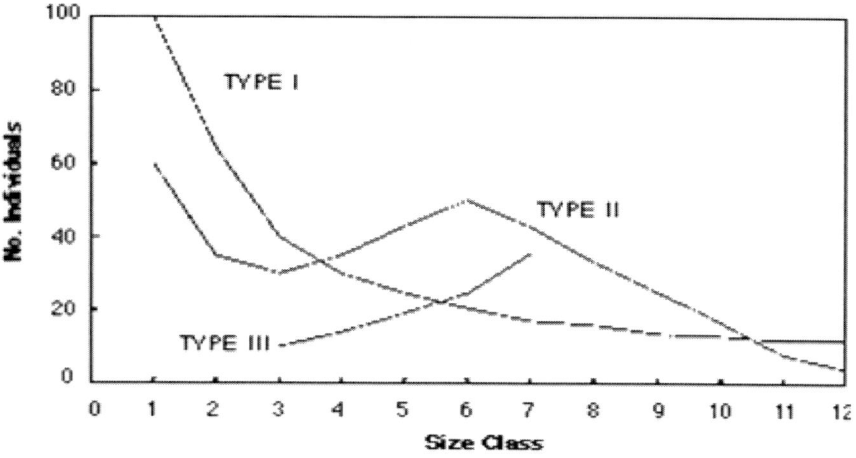

Peter, 2005.

Figure 1. Idealized size class distributions exhibited by tree populations.

Bell shaped size class (DWAF, 2005): This depicts characteristic populations which experience sporadic or irregular seedling establishment, often related to large-sized gaps in evergreen forest (DWAF, 2005; Peter, 2005). In evergreen forests, the shade-intolerant canopy species, such as *Ocotea bullata*, and invader plant species such as *Acacia melanoxylon*, typically exhibit this size class profile because they can not regenerate in shade (DWAF, 2005). Species that exhibit the bell-shaped size class profile are those that regenerate and develop in cohorts after a window of suitable conditions for regeneration do appear (Mudekwe, 2006). According to Peter (2005) the levels of regeneration for the species may be sufficient to maintain the population, but its infrequency of occurrence causes notable ''peaks'' or ''valleys'' in the size class distribution as the new seedlings grow into larger size classes. The size class profile may also imply that such a species requires a large gap to become established (DWAF, 2005). In some populations, this size class profile may imply that the regeneration is being temporarily interrupted through either excessive harvesting for timber (see Nduwamungu and Malimbwi, 1997; Mbwambo, 2000; Luoga et al. 2002; Mudekwe, 2006), fruit or seed (Peter, 2005) or direct physical damage to seedling through infrequent extreme fires (Trapnell, 1959). Or it may relate to the exclusion of fire in the woodlands that will enable fire-intolerant species to become established (Geldenhuys, 1993).

Static shape size class (DWAF, 2005): Such a size class profile reflects the population in which there is no regeneration but existing trees increase in diameter (DWAF, 2005). The regeneration of such species is severely limited for some reason (Peter, 2005). According to Chidumayo (1997), a number of limiting factors of regeneration in Miombo woodlands include fires, water and nutrient stress, insectivory, herbivory, lack of fruit and seed production. Sporadic occurrence of either of or all of these factors in the woodland or forest may result in individual species experiencing sporadic germination, and also establishment of a single or few plants at a time. Over-exploitation of mature timber species may also result in limited regeneration and consequently formation of Type III size class profile for the species that is being exploited (Nduwamungu and Malimbwi, 1997). In the absence of the conducive environment, such species may be severely limited in occurrence (Schwartz et al. 2003) or permanently disappear as this has the potential to result in decrease of the likelihood of conspecific replacement and thus increasing the risk of failure of the natural successional pathway (McKenzie, 1988).

Although the three size-class distributions correlate well with the three different regeneration guilds, it is important to remember that the population structure of a species is extremely dynamic and sensitive to changes in the level of regeneration (Peter 2005). It is important to consider that Type I is a build up of Type II (Bell shaped) curves. Each bell is a specific even-aged cohort that establishes in a specific point in time. The stems of this cohort vary in growth rate and some die, then the height of the bell decreases and the base of the bell widens. But if you have regular regeneration, different bells are integrated and then results in an inverse *J*-shaped curve (Type I).

3. CHARACTERIZING MIOMBO BASED ON STRUCTURE PROFILES AND COMMON DISTURBANCES

The miombo woodlands are a vital source of both wood and non-wood products. These provide a wide range of products for subsistence consumption and trade. Miombo woodlands contribute greatly to the livelihood security of the rural people as the rural economics is mainly based on the premises of charcoal production, shifting cultivation and to some extent timber production. Hence they are the common and major disturbances in the miombo woodland. These activities are on the increase and the result has been accelerated deforestation arising mainly from shifting cultivation and charcoal production. In order to sustainably manage the miombo woodlands, it is important to explore the potential that comes with understanding the effects of these utilization aspects and the regeneration ecology of the miombo woodlands as this has direct implication for both species persistence and for people's livelihoods. Data from three different areas that were previously under timber harvesting and undisturbed woodland stands in Zambia exhibit the bell-shaped size class profile for stems ≥ 5cm DBH (Figure 2).

This class size class profile implies that miombo woodlands experience sporadic or irregular seedling establishment. However, the size class profile of stems <5 cm DBH is the characteristic inverse *J*-shape curve (Figure 3). This means that the actual levels of regeneration in miombo woodlands may be sufficient to maintain the woodland, but its infrequency of occurrence cause noted peaks as the new seedlings grow into larger size classes. This is common in vegetation formations in which the dominant species depend on canopy gaps for regeneration. Miombo dominant species require high light intensities to develop and therefore require large gaps (Lees, 1962; Syampungani, 2008). Opening up of miombo canopy stimulates germination of the soil seed reserves as the temperature and light intensity reaching the woodland floor increase. Exposure of stumps to sunlight enhances coppicing effectiveness of most of miombo woodland species (Handavu et al. 2011). Most miombo woodland species are capable of developing vegetatively because they generally have both extensive and horizontal root systems which facilitate rapid recuperation after cutting (Malimbwi et al. 1994).

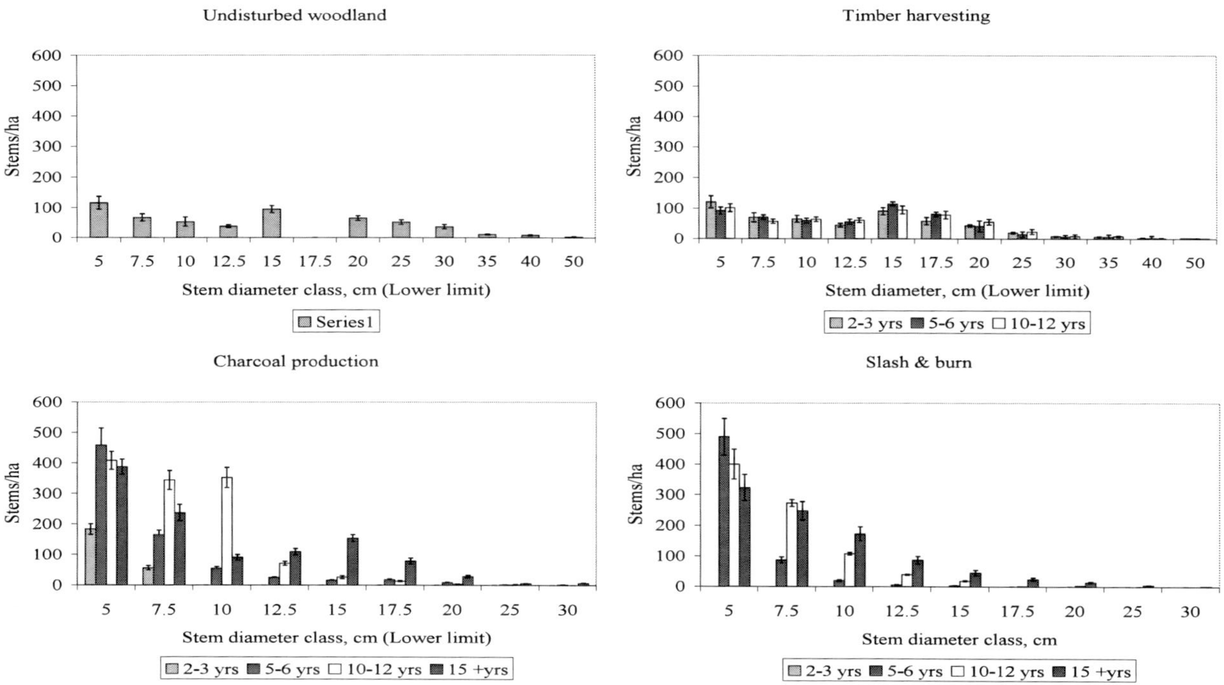

Source Syampungani, 2008.

Figure 2. Stem diameter distribution of stems ≥ 5cm DBH in stands previously under different disturbances and the relatively undisturbed woodland stands.

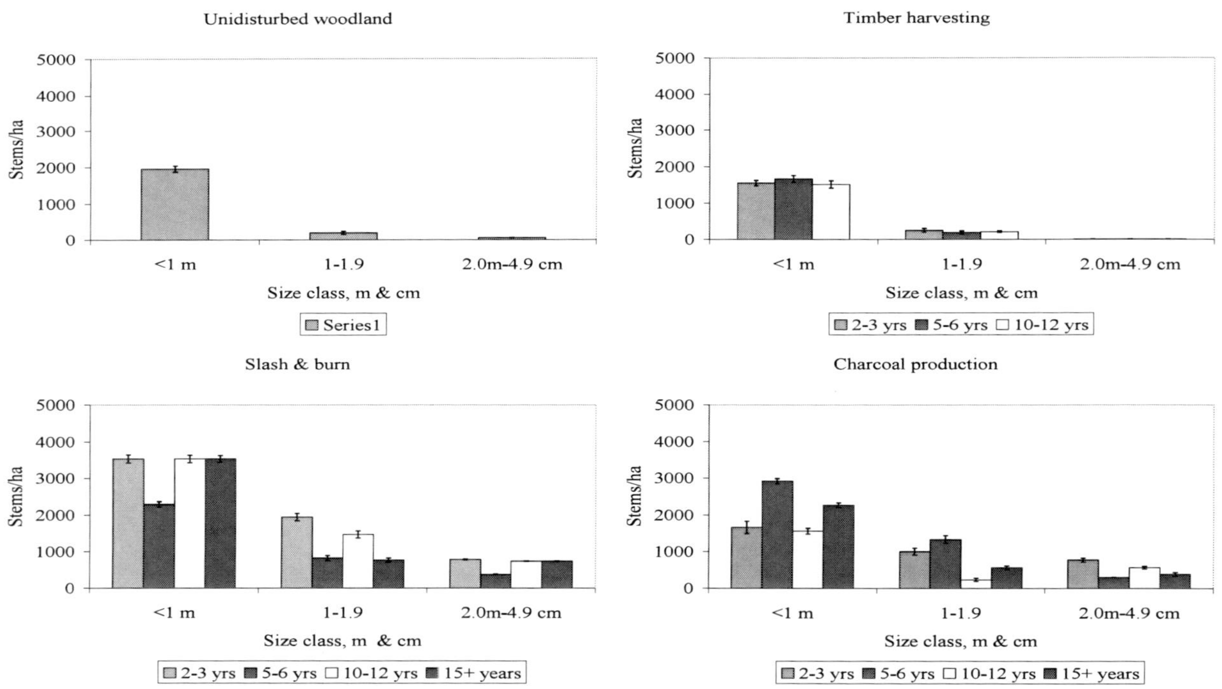

Source Syampungani, 2008.

Figure 3. Stem size distribution of woody stems <5 cm DBH in woodland stands previously under different disturbances and the relatively undisturbed stand.

3.1. Silvicultural Systems and Key Economic and Ecological Trees of Miombo Woodland

The key economic and ecological miombo woodland species are Brachystegia boehmii, Brachystegia floribunda, Brachystegia longifolia, Brachystegia spiciformis, Julbernadia paniculata and Albizia antunesiana. These are dominant and co-dominant species of miombo woodland and are used for timber, charcoal production and also as ash fertilizers by shifting cultivators. The rural livelihood of eastern and southern Africa mostly depend on forests including slash and and burn agriculture, charcoal and timber production. Each of these utilization aspects has its own impact on the forests and woodlands in general. Charcoal production and slash and burn agriculture have been associated with massive loss of forests and biodiversity in many parts of the miombo ecoregion. Up to 400, 000 ha $year^{-1}$ of forests have been reported deforested as a result of charcoal production and slash and burn agriculture in the miombo ecoregion (Syampungani, 2008 and literature cited therein). Charcoal production and slash and burn agriculture have time immemorial been blamed as the principal causes of deforestation and its associated negative impacts in the tropical Africa (Jepma, 1995). Timber production which is mainly through single tree selection harvesting has its own impact not on the forest as a whole but on individual species that are harvested. Various researchers in many parts of the miombo ecoregion namely; Zambia (Syampungani, 2008); Tanzania (Luoga et al 2002); Malawi (Makungwa and Kayambazinthu, 1999) and in Mozambique (Grundy and Cruz, 2001) have observed low levels of species such as *Pterocrapus angolensis, Albizia antunesiana, Brachystegia floribunda* etc that were being harvested for timber production. This has resulted in what Schwartz et al (2002) referred to as economic extinction of species. The economic extinction allows such species to persist at low densities which make their exploitation costly. This is also an indication of unstable population that tends to decrease the likelihood of conspecific replacement and increases the risk of collapse of the natural successional pathways. According to McKenzie (1986) the collapse of natural pathwy has the ability to result in total collapse of the entire population if such a situation persists for a long time as there would be no replacement of individual plants. Shackleton et al (2005) indicated that there is actually a growing evidence and concern that levels of extraction for timber, carving and poles are unsustainable for many species in Southern Africa. This may be attributed to the fact that most of the species harvested are mostly dominant species and therefore are light demanding. Single tree harvesting does not provide for maximum light for the seedling to recruit into high diameter classes. However, studies have also shown high regeneration of these species in areas that were under slash and burn agriculture and charcoal production (Luoga et al 2002; Syampungani, 2008).

It is therefore possible that a silvicultural system that integrates charcoal production, slash and burn agriculture and timber harvesting can be developed. This would involve zoning the forest resource area into management classes. Since both slash and burn and charcoal production result in enhanced seedling/recruitment establishment of timber species on termination of the disturbance impacts over an area, their regrowth stands may be managed for selected trees of timber to grow into good-sized trees. Management should involve appropriate silviculture for improved quality of potential logs from young trees in those stands through pruning. Additionally, thinning should also be carried out to enhance the growth of the intended crop. During thinning, the thinned out stems can be used as poles for construction. Other silvicultural operations that should be carried out should include fire prevention and drought impact reduction through weeding around seedlings. Additionally, the reduction in stocking may also reduce water stress arising from competition. Management of coppices to ensure high survival of plants should be encouraged. This should begin with the protection of stumps against fires and herbivory to enhance the survival rate of coppices. Secondly, increased stump heights during felling as observed by Grundy (1990) can enhance the survival of stumps and coppicing. Adhering to optimum diameter classes (see Handavu et al 2011) within which particular species have high coppicing effectiveness can enhance coppicing of such species. Lastly, other factors that should be considered to enhance coppice effectiveness include plant age, stump height and surface area.

3.2. Conclusion

Miombo woodland is central to rural livelihoods of many people of Southern Africa as the rural economy mainly depend on products from the woodlands. The exploitation of miombo woodland especially for charcoal production and shifting cultivation has been observed to have serious negative implications in the many parts of Southern Africa. However, there is enough scientific evidence indicating that the woodland recovers upon cessation of either charcoal production or shifting cultivation. Additionally, scientific evidence also shows that suggests that timber species such as *Pterocarpus angolensis, Albizia antunesiana, Brachystegia floribunda* etc do better in opened up areas than in forested areas when harvested for timber production. This evidence suggests that timber production (single tree selection harvesting) could go side by side with either charcoal production or shifting cultivation to enhance the survival of regeneration for these species. It further gives an indication of the current silvicultural systems could be designing to include these disturbance factors in the management of miombo woodland.

REFERENCES

Brigham, T., Chihongo, A., Chidumayo, E. 1996. Trade in woodland products from the Miombo ecoregion. In: Campbell, B. (Ed.), The Miombo in Transition: Woodland and Welfare in Africa. *CIFOR, Bogor.* pp.137-174.

Chidumayo, E.N. 1997. Miombo ecology and management: An introduction. Stockholm Environment Institute, Stockholm.

DWAF. 2005. Sustainable resource use. PFM Guideline for Department of Water Affairs and Forestry. DWAF, Ramboll and Danida, Pretoria.

Geldenhuys, C. J. 1993. The Use of Diameter Distributions in Sustained-Use Management of Forests: Examples from Southern Africa. In: Piearce, G.D., Gumbo, D.J. (Ed.), The Ecology and Management of Indigenous Forests in Southern Africa: Proceedings of an International Symposium, Victoria Falls, Zimbabwe, 27-29 July 1992. Zimbabwe Forestry Commission and SAREC. pp. 154-167.

Geldenhuys, C.J. 2010. Managing forest complexity through application of disturbance-recovery knowledge in development of silvicultural systems and ecological rehabilitation in natural systems in Africa. *Journal of Forest Research* 15(1): 3-13.

Grundy, M.I., Cruz, R.M. 2001. Impact of Harvesting for Building Poles on the Woodland Structure of Derre Forest, Mozambique. In: Matakala, P., Sitoe, A (Eds.), Proceedings of the international workshop on Miombo woodlands in the New Millenium: Trends, uses and their Role in Sustainable Development, Casa Msika, Manica, Mozambique, September 26-28. Pp.78-96.

Handavu, F., Syampungani, S., Chisanga, E. 2011. The influence of stump diameter and height on coppicing ability of selected key Miombo woodland tree species of Zambia: A guide for harvesting for charcoal production. *Journal of Ecology and the Natural Environment* 3(14): 461-468.

ITTO 1992. ITTO guidelines for sustainable management of natural tropical forests. Technical Series No. 5, International Tropical Timber Organisation, Yokohama.

Jepma, C.J. 1995. Tropical Deforestation: A Socio-Economic Approach. Earthscan, London.

Lees, H.M.N. 1962. Working plan for the forests supplying the Copperbelt Western Province. Government Printer, Lusaka.

Luoga, E.J., Witkowski, E.T.F., Balkwill, K. 2002. Harvested and standing wood stocks in protected and communal woodlands of eastern Tanzania. *Forest Ecology and Management* 164: 15-30.

Makungwa, S., Kayambazinthu, D. 1999. Monitoring and evaluating productivity in miombo woodlands under co-managements: The case of Liwonde Forest Reserve. In: Ngulube, R.M., Mwabumba, L., Chirwa, P.W. (Eds.), Proceedings of a National workshop, Sun and Sand Holiday Resort, Mangochi, Malawi, September, 27-29, pp. 115-130.

Malimbwi, R.E., Solberg, B., Luoga, E.J. 1994. Estimation of biomass and volume in Miombo woodland at Kitulanghalo Forest Reserve, Tanzania. *Journal of Tropical Forest Science* 7: 230-242.

Mbwambo, L.R. 2000. Species utilization and resource potential of Miombo woodlands: A case of selected villages in Tabora, Tanzania. Msc. Thesis, University of Stellenbosch.

McKenzie, B. 1988. Guidelines for sustained use of indigenous forests and forest products. Occasional report number 35, Foundation for Research and Development, Pretoria.

Mudekwe, J. 2006. Demography and dynamics of commonly harvested woody species in a protected indigenous forest in Zimbabwe. In: Bester, J.J., Seydack, A.H.W., Vorster, T., van der Merwe, I.J., Dzivhani, S (Eds.), Multiple Use Management of Natural Forests and Woodlands: Policy Refinements and Scientific Progress. Proceedings of a Natural Forests and Savanna Woodlands Symposium IV, Port Elizabeth, South Africa, 15-18 May 2006. pp.183-197.

Nduwamungu, J., Malimbwi, R.E. 1997. Tree and shrub diversity in Miombo woodlands: a case study of kitulanghalo Forest Reserve, Morogoro, Tanzania. In: Imana-Encinas, J., Klein, C (Eds.), Proceedings of the International Symposium on Assessment and Monitoring of Forests and in Tropical Dry Regions with Specific Reference to Gallery Forests. University of Brazil, Brazil, pp. 239-258.

Peter, M.C. 2005. Sustainable Harvest of Non-timber plant resources in tropical moist forest: An Ecological primer. *http:/www.worldwildlife.org*. Date visited: 12/11/12.

Schwartz, M.W., Caro, T. M., 2002. Effect of selective logging on tree and understory regeneration in miombo woodland in western Tanzania. *African Journal of Ecology* 41: 75-82.

Syampungani, S. 2008. Vegetation Change Analysis and Ecological Recovery of the Copperbelt Miombo woodlands of Zambia. PhD Thesis, University of Stellenbosch, RSA.

Trapnell, C.G. 1959. Ecological results of woodland burning experiments in Northern Rhodesia. *Journal of Ecology* 47: 129-168.

INDEX

D

E

F

I

J

K

L

M

Q

R

S

T

U

V

W

X

Y

Z